Springer Series in Materials Science

Volume 170

For further volumes:
http://www.springer.com/series/856

The Springer Series in Materials Science covers the complete spectrum of materials physics, including fundamental principles, physical properties, materials theory and design. Recognizing the increasing importance of materials science in future device technologies, the book titles in this series reflect the state-of-the-art in understanding and controlling the structure and properties of all important classes of materials.

Anatoly Lanin

Nuclear Rocket Engine Reactor

Springer

Anatoly Lanin
St. Petersburg
Russia

ISSN 0933-033X
ISBN 978-3-642-32429-1 ISBN 978-3-642-32430-7 (eBook)
DOI 10.1007/978-3-642-32430-7
Springer Heidelberg New York Dordrecht London

Library of Congress Control Number: 2012949701

Printed on acid-free paper

Springer is part of Springer Science+Business Media (www.springer.com)

Preface

This book covers the problems of theory and practice of developing core elements for a high-temperature Nuclear Rocket Engine reactor (NRER). As a concept, the problem is solved through a non-separable interrelation of the structure, properties and performance patterns of the structural elements. Due to high brittleness of materials used in a structure, on the basis of interstitial phases, i.e., carbides, nitrides, hydrides, special focus is put onto reasoning stability of the core elements being exposed to high heat and neutron flux, high stresses, aggressive gases, and ultra-high temperature. Such a variety of performance parameters required development of a whole range of new pre-reactor and reactor methods of material testing and execution of a complex investigation of the basics for the structure formation of these materials, in order to ensure the maximal reliability of the structures operation. Development of Fuel Elements (FE), simple by their shape, but operating under extremely heavy conditions, required involvement of the experts in various sciences, such as physics, mechanics, heat engineering, reactor and radiation material science, production technologies of nuclear fuel, and structural materials. The developed phenomenological and statistical-thermodynamic theories of three-component interstitial alloys and the set of laws for modification of diffusion-controllable processes (creep, sintering, carbonizations, and corrosion) have deepened present scientific presentation of these processes and have got wide recognition in the scientific community. Doubtless interest is presented also on irradiation behavior of carbide materials and bearing capacity of fuel and constructional materials on the basis of interstitial phases in conditions of an irradiation neutron fluxes.

There are numerous separate works discussing development and production of FEs for thermal neutron and fast neutron reactors, but there is no book covering a problem of developing core elements for Fuel Assemblies (FA) integrally. The fact, that results of exploring on the NRE reactor are disseminated under various publications of periodic journals or express collections, access to which is complicated and own experience are great, so it became the incentive reason for writing this monograph in which data on a resolving of a problem of FA creation for high-temperature solid phase NRE reactor, characteristics of materials and principles of designing are generalized. Author realizes that the frame of concerned subjects in a book is restricted. Many problems of design, neutron-physical and hydro dynamical characteristics of the NRE reactor are handled in the book *Nuclear rocket engines* [1].

Many scientific investigators and engineer's of the Research Institute of Production Research Association (RIPRA) "Luch" participated in a development, manufacturing and material tests of the NRE are presented in this book [2]. The circle of authors shared in creation of the NRE is rather wide and, certainly, is not terminated by the authors presented in the bibliographic list of this book. Many actively worked authors who died, did not publish in due time their manuscripts in the open printing. So it should note the big contribution of Ph.D. N. Poltoratsky and Doctor of Science L. B. Nezhevenko in the development of the fuel elements technology HRA from carbide and carbide-graphite materials and Ph.D. A. L. Epstein for the material science organization and technological works on neutron moderator from hydrides of zirconium and yttrium. The author remembers the creative contacts with them with gratitude, and also a fruitful cooperation with many workers with whom he had the luck to work at the NRE elaboration in scientific research institute of RIPRA "LUCH" since 1962 year.

The author expresses gratitude to Professor R. A. Andrievsky for the review of the manuscript and useful advices and to engineer A. P. Abramova for the book formatting.

St. Petersburg Anatoly Lanin

References

1. Demjanko, J. G., Konuhov, G. B., Koroteev, A. S., Kuzmin E. P.,& Pavelev, A. A. (2001). *Nuclear rocket engines* (p. 413). Moscow: Open Company "Norm-inform".
2. RIPRA "Luch". (2004). Affairs and people. Podolsk. RIPRA "LUCH". Ed. Fedik I. I. Podolsk, 455 p.

Contents

Acronyms

AEI	Atomic Energy Institute by Name I. V. Kurchatov
AIAM	All Union Institute of Aviation Materials
ARIIM	All Union Research Institute of Inorganic Materials by Name A. A. Bochvar
BG	Bearing Grid
BNU	Bearing Nozzle Unit
DLCA	Development Laboratory of Chemical Automation
ETC	Experimental Technological Canal
FA	Fuel Assembly
HIP	Heat Insulating Packet
HRA	Heat Releasing Assembly
HS	Heating Sections
IHT	Institute of High Temperature
ITER	International Thermonuclear Experimental Reactor
IVG-1	Research high temperature gas cooled reactor IVG-1
LDP	Low Density Pyrographite
LRE	Liquid Rocket Engine
NRER	Nuclear Rocket Engine Reactor
NFA	Nuclear Fuel Assemble
NPI	Nuclear Power Installation
NEPI	Nuclear Engine Power Installation
PEI	Physical Energy Institute
PGV	Pyrographite
RDIET	Research and Development Institute of Energy Technology by Name N. A. Dollezhal
RIHRE	Research Institute of Heat Releasing Elements "Luch"
RIPRA	Research Institute of Production Research Association "Luch"
RITM	Research Institute of Test Machine
RITP	Research Institute of Thermal Processes by Name V. M. Ievlev
SIAC	State Institute of Applied Chemistry

TC	Technological Canal
TEC	Thermoelectric Converter
TECs	Thermoelectric Converters
TEND	Thermal-Electric Neutron Detectors
TSR	Thermal Stress Resistance

Chapter 1
History Creation of a Nuclear Rocket Engine Reactor

The high pace of development and the high scientific and technological level of atomic power engineering achieved in the USSR in the second half of the 1950s created objective prerequisites for the construction of fundamentally new types of stationary and mobile nuclear energy reactors and facilities intended, in particular, for cosmic purposes, such as nuclear rocket engines (NREs) and small-size nuclear–electric energy converters [1, 2].

The development of NREs in the USSR was initiated in 1955 when I. V. Kurchatov, S. P. Korolev, and M. V. Keldysh met to discuss the possibility of creating a rocket with an atomic engine to enhance the defensive power of the country in response to the Rover program for developing NREs, started in the USA (Fig. 1.1).

In 1956 and 1958, two government acts for the development of work with the aim of building nuclear rocket engines were signed in the USSR [2]. At this stage, preliminary design work was started to create a mounting base for testing NREs, and material technology investigations were initiated. The heat exchange and hydrodynamic conditions were studied at Research Institute-1 [currently, Research Institute of Thermal Processes by V. M. Ievlev (RITP)]. Studies in the field of neutron physics and reactor control were performed at the Obninsk Physical Energy Institute (PEI) and at the Kurchatov' Atomic Energy Institute (AEI). The technology of refractory materials and heat-releasing elements (HREs) was developed at Research Institute-9 (currently, the A. A. Bochvar All-Union Research Institute of Inorganic Materials, ARIIM). The problems of selecting materials for NREs based on carbides of transition metals and graphite in gas media were also partially investigated, beginning from the 1960s, at the All-Union Institute of Aviation Materials (AIAM), the State Institute of Applied Chemistry (SIAC), the Institute of High Temperatures (IHT), the Graphite Research Institute, and the Ceramics Department of the Leningrad Technological Institute.

In 1957, researchers at the RITP proposed a principle of refining individual elements of units of the NRE reactor on electrothermal and plasmatron mounts, which reduced the extent of reactor tests. Different NRE schemes were considered. The simplest NRE with a reactor with a solid-state core (named the A scheme) can produce

A. Lanin, *Nuclear Rocket Engine Reactor*, Springer Series in Materials Science 170,
DOI: 10.1007/978-3-642-32430-7_1, © Springer-Verlag Berlin Heidelberg 2013

Fig. 1.1 From *left* to *right*: Academicians of Science academy of the USSR, the main designer of ballistic missiles S. P. Korolev, the main supervisor of a nuclear studies problem I. V. Kurchatov, president of Science academy of the USSR M. V. Keldysh

a thrust 450 s. The NRE with a gas-phase reactor (the B scheme) could produce a specific thrust up to 2,000 s, but problems to be solved in its construction, namely, the confinement of a uranium plasma with the temperature up walls of a heat-releasing assembly (HRA) from the plasma, were much more severe than those in the construction of the A scheme. It was decided to begin the development of the A scheme (while the work related to the B scheme was continued as a research study).

Within 2 years, two teams were actively involved in the development of the first NREs. A ground-based IVG-1 (research high-temperature gas-cooled) reactor, a prototype of an in-flight NRE version, was developed at the Research and Development Institute of Energy Technology (RDIET) headed by N. A. Dollezhal' and at AEI under the supervision of N. N. Ponomarev-Stepnoi. The second team developing a ground-based IR-100 engine was headed by V. M. Ievlev, the scientific chief of RITP, A. D. Konopatov, the chief designer of the engine at the Development Laboratory of Chemical Automation (DLCA), Voronezh, and V. Ya. Pupko, the scientific chief of the reactor at the Physical Energy Institute (PEI), Obninsk. The first carbide-based HREs were proposed and manufactured at Research Institute-9. The expansion of NRE studies required the manufacture of a great number of HREs for experimental HRAs. In August 1962, the Ministry of Medium Machine Building of the USSR decided to create the Research Institute of Heat Releasing Elements (RIHRE) [currently, the "Luch" Research and Production Association (RPA), a Federal State Unitary Enterprise (FSUE)] with an experimental plant providing technological possibilities for the rapid complex development and production of new types of nuclear fuel and HREs [2]. This resolution was preceded by the very important decision of 30 December 1959 about the inexpediency of further developments of military nuclear ballistic rockets (due to progress in the development of chemical-fuel engines) and the necessity of continuing NRE developments for space launchers [1].

The basic challenge encountered by the developers of NREs was to prevent damage to the ceramic active core and, most of all, HREs caused by thermal stresses. It is known that thermal stresses are proportional to the product of the energy-release density, the elastic modulus of the material, the linear expansion coefficient, and the square of the characteristic transverse size of the construction. The energy release cannot be strongly reduced because such a reduction would result in an increase in the reactor weight with protection and would reduce the NRE advantages to a minimum. Reducing the HRE characteristic size can be done efficiently, but this is restricted by the fact that elements of very small sizes would be severely damaged by vibrational stresses, which are significant in rocket engines. In this case, the limiting characteristic transverse size of HREs is 2–3 mm. An alternative to the carbide HRE is the graphite HRE. Graphite has a unique thermal strength because its elastic modulus is almost two orders of magnitude smaller than that of carbides. However, a substantial disadvantage of graphite is that it quite actively interacts with hydrogen (everything comes at a price).

Unlike American scientists, Soviet researchers began to develop the NRE core using HREs made not of graphite, which is thermally stable but unstable in the hydrogen medium, but of carbides, which are brittle but are more stable in hydrogen [2]. While in many other scientific fields Soviet scientists had to 'overtake' foreign investigations, the NRE was developed without blindly copying 'foreign' samples, the experience of predecessors being critically analyzed. Soviet researchers decided (correctly, as the result showed) that protecting HREs from interaction with hydrogen is more difficult than providing its acceptable thermal strength resistance. Although graphite HREs was also developed in the USSR, they were regarded as a backup.

In 1962–1969, under the supervision of M. V. Yakutovich, director of the RIPRA, the material technology, technological, theoretical, and test departments were created to solve the following problems [2]:

- selection of the fuel and construction materials for NREs;
- development of the technological foundations for manufacturing devices from these materials;
- construction and computational and experimental demonstration of the efficiency of elements of the active core; preparation of project documentation and organization of semicommercial production.

These problems were successfully solved by using a systematic and complex approach. Detailed theoretical estimates of the operational conditions of the devices, the required technological studies of the properties of materials, and the working out of technological versions were realized in the form of technological chains developed together with the experimental plant. In turn, experimental and test samples of devices were subjected to thermal, hydraulic, and resource tests in the IVG reactor located on the nuclear proving ground "BAIKAL-1" in the Semipalatinsk region (town Semipalatinsk-21, later Kurchatov, Kazakhstan). The working area "BAIKAL-1"

Fig. 1.2 Director of the RIPRA "LUCH", Doctor of Science, laureate of many state prize M. V. Yakutovich (10.08.1902–29.06.1988)

is situated at 70 km from inhabited town Kurchatov and at 150 km from the Semipalatinsk town (Figs. 1.2–1.4).

This paper is devoted to the history of material technology developments for manufacturing elements for the NRE reactor core at Research institute of Production Research Association [RIPRA] Federal State Unitary Enterprise (FSUE) "Luch" from 1962 to 1991. Because the results of NRE studies are scattered over numerous publications in periodicals or special collections of papers that are not easily accessible, we decided to generalize the data on the characteristics of the materials and the efficiency of elements of the NRE core in this paper.

Clearly, the scope of problems considered here is limited. For example, we do not discuss issues concerning the design and the radiation and thermal protection of the nuclear engine, or the neutron-physical and hydrodynamic characteristics of the reactor. These issues were partially considered in the book Nuclear Rocket Engines published in 2001 [1].

Because of the high brittleness of the materials based on interstitial phases (carbides and hydrides), which are used in construction, special attention is paid in manufacturing active-core elements that are stable under the action of high thermal and neutron fluxes, stresses, aggressive gas media, and ultrahigh temperatures. Such a variety of operating parameters required the development of a number of new prereactor physic-mechanical methods for testing materials at the RIPRA "Luch", where diffusion-controlled processes (creep, corrosion, and radiation) were investigated in fuel and construction materials based on interstitial phases.

Beginning from 1976, HRAs developed for NREs were tested in the IVG-1 reactor, which was used to work out the elements of the reactor core. Later, a mount was constructed for testing an 'engine' version of the IR-100 reactor, which was tested for several years at different powers and was then converted into a material technology research low-power reactor, which is still working successfully till date.

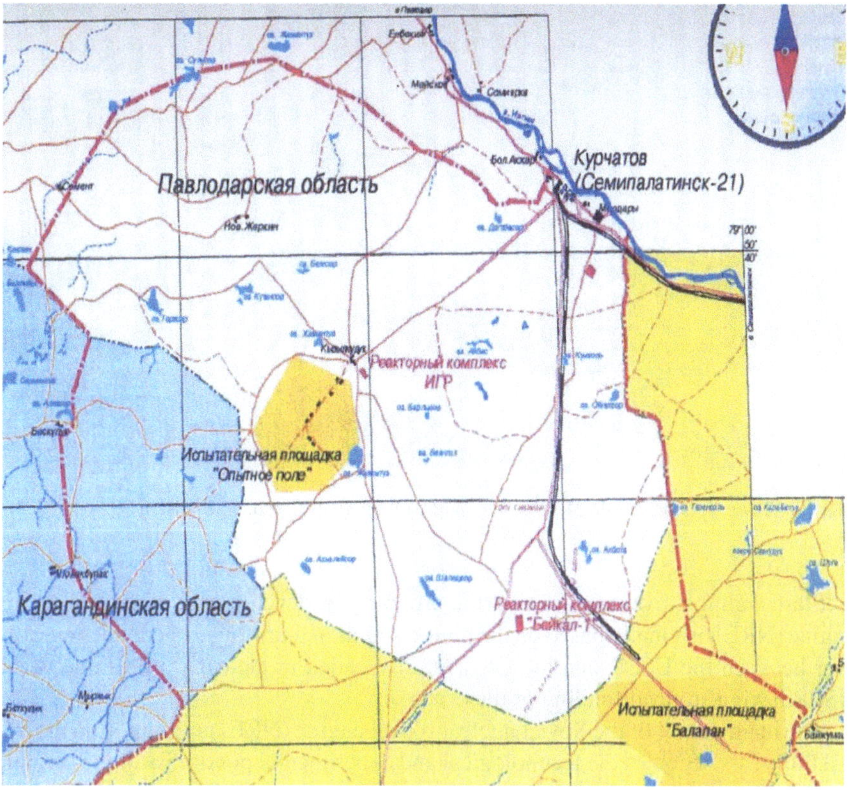

Fig. 1.3 The map of the Semipalatinsk region (Kazakhstan)

In the late 1970s and early 1980s RITP and RDIET have commenced an intensive work on developing a multimodal system [3] capable of producing both jet thrust and electricity to power life support systems of the spacecraft. Besides the main nuclear propulsion mode, the NRE was to be operating at two generation modes: low-power mode for prolonged operation (several years), and high-power mode for half of the specified service life in the propulsion mode. The high-power mode (HPM) presented no particular problems for the reactor. At low-power mode (LPM), the heat transfer agent circulates only outside the NFA casing, while the heat from fuel rods is transferred to the casing by radiation through the thermal insulation. Such mode differs significantly from the propulsion mode, the former involving considerable temperature gradient across the NFA radius and uranium burnout (min. 3–5 %). Therefore, the applicability of the structural NFA parts and fuel rods under these conditions demands further research. First of all, the design and processing technology of the fuel rods should guarantee retention of fission products inside the rod for several years at temperatures of 2,000 K under high vacuum or in H-containing working fluids at pressures of 0.1–0.2 bars.

Fig. 1.4 Town Semipalatinsk-21, later Kurchatov on the river Irtysh, Kazakhstan

Started almost five decades ago, the program for development of nuclear rocket engine (NRE) originally based on the political aims and priorities of conducting Cold War between the USSR and the US were suspended in the early 1990s due to the USSR having stopped funding of these works.

The main output of the Soviet nuclear rocket engine (NRE) program from the initial 60x years becomes the technological and experimental possibility of the ceramic active zone creation of the NRE for hydrogen heating at the maximum parameters values: A hydrogen reheat at temperature up to 3,100 K; power fuel density to 35 MW/l; duration of tests to 4,000 s; the maximum heating/cooling rates of fuel elements to 400/1,000 K/s

The results of numerous studies performed at the RIPRA "Luch" for many years allowed constructing a scientific system of knowledge about the nature of refractory compounds, mechanisms of their deformation and damage, and the features of their behavior in construction and fuel hardware during variation of the operational parameters of reactors. Many developments at the RIPRA "Luch" have been acknowledged at international conferences, symposia, and exhibitions. A group of scientists has grown who have defended more than 30 Doctor of Science and 200 Ph.D. dissertations and published a few dozen monographs and a few hundred scientific papers on high-temperature materials and their applications. Some researchers became laureates of State Prizes of the USSR and Government Prizes and received titles of Honored Scientist, Honored Technologist, and Honored Inventor.

The workings out of NRE begun almost half a century ago caused, in due time, by political problems and priorities in conducting "cold war" between the USSR and the USA were suspended in the early 1990s in connection with the financing termination in the USSR.

References

1. Demjanko, J. G., Konuhov, G. B., Koroteev, A. S., Kuzmin E. P., & Pavelev, A. A. (2001). *Nuclear rocket engines* (p. 413). Moscow: Open Company "Norm-inform".
2. RIPRA "Luch". (2004). Affairs and people. Podolsk. RIPRA "LUCH". Ed. Fedik I. I. Podolsk, 455 p.
3. Ponomarev-Stepnoy, N. N. (1993). Creation history of NRER in the USSR (pp. 3–18). *Third Branch Conferences "Nuclear Power in Space"*.

Chapter 2
Design of a NRER

The concept of a nuclear rocket engine is simple and consists of using a nuclear reactor instead of a combustion chamber for heating gas producing the thrust during its escape from a supersonic nozzle. The efficiency of a rocket engine is determined by the ejection rate of gas from the nozzle, which is inversely proportional to the square root of the molecular weight of the gas. Hydrogen has the smallest weight (2 a.m.u.). The ejection rate of hydrogen heated up to 3,000 K is more than twice that of the best chemical fuels for LREs. This is an advantage of the NRE, which can heat pure hydrogen (the mean molecular weight of combustion products in LREs always exceeds 10 a.m.u.). Instead of the ejection rate, the engine efficiency is often characterized by the specific thrust, equal to the ratio of the momentum imparted to the rocket engine to the mass flow rate of the working substance. (The specific thrust is also defined as the ratio of the thrust to the weight flow rate of the fuel and is measured in this case in seconds.) According to calculations, the mass delivered from a circum-terrestrial orbit to the geostationary orbit by an interorbital spacecraft equipped with an NRE having a specific thrust in the range from 850 to 4,400 s is three times larger than the mass delivered by a spacecraft with an LRE.

Aside from these important advantages, the NRE has substantial disadvantages. The main disadvantages are as follows. First, the NRE reactor is considerably heavier than the LRE combustion chamber. Second, the reactor is a high-power radiation source requiring a radiation shield. This makes the engine even heavier and considerably complicates its operation. The requirement of the stability in hydrogen at high temperatures and neutron-physics restrictions strongly reduce the choice of materials that can be used for manufacturing HREs and high-temperature elements of the HRA. Because the NRE is an air-borne reactor, it should be light enough. Therefore, this reactor should provide a very high energy density, exceeding the energy density of ground energy reactors by a few orders of magnitude [1, 2].

A number of important principles for using brittle carbide materials in HRAs were formulated during the construction of the HRA. The HRA should be made of functionally autonomous units and remain operable even if some of its units are damaged. Each unit is made as a technologically independent aggregate that does

A. Lanin, *Nuclear Rocket Engine Reactor*, Springer Series in Materials Science 170,
DOI: 10.1007/978-3-642-32430-7_2,

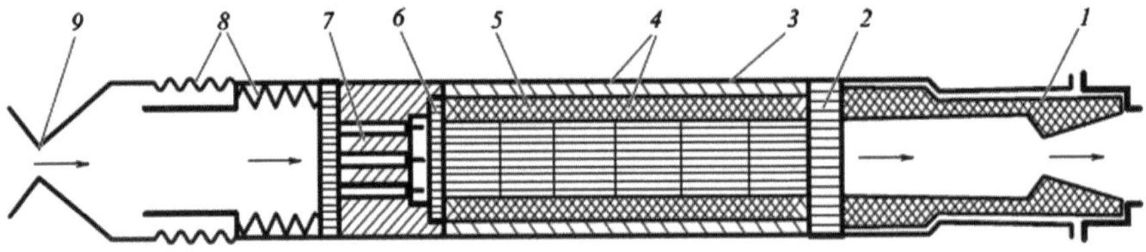

Fig. 2.1 Principal HRA scheme: *1* nozzle unit; *2* bearing grid; *3* beryllium-steel bilayer housing; *4* heat-insulating packet; *5* heating units; *6* input grid; *7* end deflector; *8* thermal-expansion compensation unit; *9* throttle

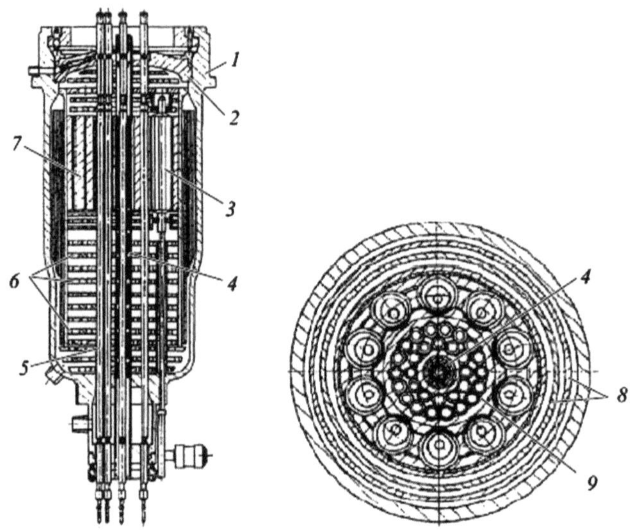

Fig. 2.2 Longitudinal and transverse sections of the IVG-1 reactor

not require complicated connections with adjacent units during its mounting into the assembly.

Such a functional, technological assembly of HRAs considerably reduced the time of experimental studies, reduced the cost of the unit, unified the unit manufacturing quality control, improved the prediction of its efficiency, and provided the maximal stability with respect to this load. The basic construction of the first-generation HRA in a heterogeneous NRE is shown in Fig. 2.1. In the upper part of HRA housing is placed an ordered set of ceramic elements that are not attached to each other. The lower part of HRA developed at the RDIET contains input channels of a working medium (hydrogen or nitrogen at a stage of cold purges).

The bench IVG-1 reactor, constructed to work out the HRA design for the NRE operation parameters [2], is a heterogeneous gas-cooled reactor with a water moderator and a beryllium deflector (Fig. 2.2). It consists of stationary and removable parts. The stationary part includes housing 1 of the reactor with lid 2, deflector 7, barrels 3 for power control, biological protection units 6, and screens 8.

Fig. 2.3 The nuclear ground "BAIKAL-1" removing a cover from the IVG-1 reactor before its nuclear start

The removable part of the reactor core contains central assembly 9 with a set of 30 technological channels (TC_S) 5 and central channel 4. The HRAs under study can be placed in both the TC group and the central channel, where the thermal neutron flux can be approximately doubled compared to its cross-section averaged value owing to a beryllium reflector surrounding the channel, which allows testing HRAs mounted in the central channel at forced (up to damaging) loads. The use of water in the bench NRE prototype instead of a hydride-zirconium moderator, which is close in its nuclear physical properties to zirconium hydride, expands the experimental possibilities of the reactor, allowing the replacement of units under study without constructive finishing, and improving the reliability of the reactor operation (Fig. 2.3).

Heat releasing elements are located in the HRA heating unit 29.7 mm in diameter and 600 mm in length (Fig. 2.4a). The heating section (HS) is divided into 6 heating parts, each of which contains 151 twisted-fuel elements and 12 semi-cylindrical fillers. Fuel elements and fillers are close-packed into a triangle lattice. The FEs is twisted along the axis. The relative diameter of the FE is 2.2 mm, the blade thickness is 1.24 mm, and the twisting step is 30 mm (Fig. 2.4b).

The specific heat release and temperature in the FEs along the HRA length are distributed no uniformly, with a maximum at the central part of the HRA in the third heating section, while the temperature of the hydrogen medium monotonically increases toward the nozzle output (Fig. 2.5).

Heating sections are intended for heating the working substance up to a specified temperature. The first four HSs, counting from the entrance of the working medium into the HRA, consist of FEs made of a double solution of carbide-graphite's; the last two of the HSs contain FEs made of a triple ($ZrC + NbC + UC$) solution. Each

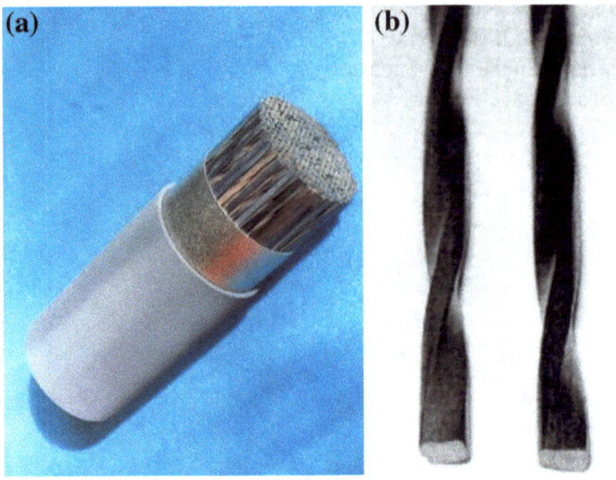

Fig. 2.4 Photo of a heating generation unit (**a**) and twisted two-blade fuel element 2.2 mm in diameter (**b**) [3]

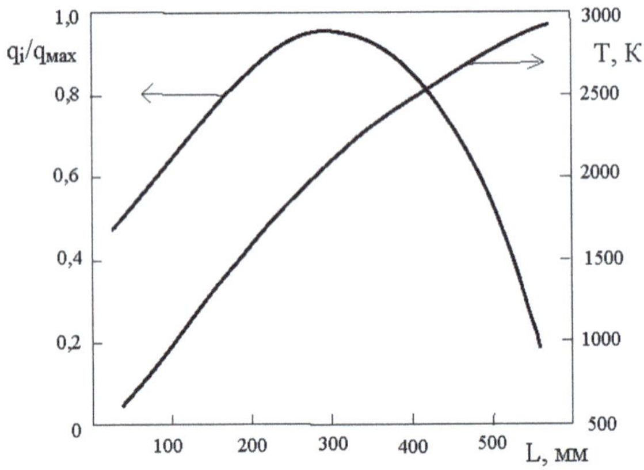

Fig. 2.5 Variation of the heat release q_i/q_{max} and a temperature along the HRA length [2, 4]

section is assembled from FEs of three types containing uranium in different amounts and located in three zones. Profiling by the uranium load was performed to level the temperature field over the HS cross-section [4]. Formulas for calculation of principal stresses σ_1 and σ_2 is of cross-section double-blade fuel elements in a stationary regime were obtained by numerical method.

Value of numerical factor matches to conditions of heat exchange Bi $= 0.25$, value in brackets for Bi $= 3.5$ (involve a change range of these parameters); the Poisson's ratio ν is equal to 0.2, α-a coefficient of linear expansion, E—is Young modulus, Bi $= \alpha_t D/\lambda$, α_t is convective heat exchange factor, and D—the fuel element diameter.

The thermal stress changes essentially at high temperatures at the expense of stress relaxation. Thermal stresses relax especially strongly in fuel elements and casings. Calculations of a relaxation usually consider only an unsteady creep presented by experimental dependence: $d\varepsilon_x/dt = B(t)\sigma_{xm}$.

Remark It is seen, that tensile stresses $\sigma_{\varphi\varphi}$ and σ_{zz}, on a cooled surface of fuel element (Table 2.1) are not equal and their values with maximum in two points

Table 2.1 Formulas for calculation of principal stresses σ_1 and σ_2 is of cross-section double-blade fuel elements in a stationary regime

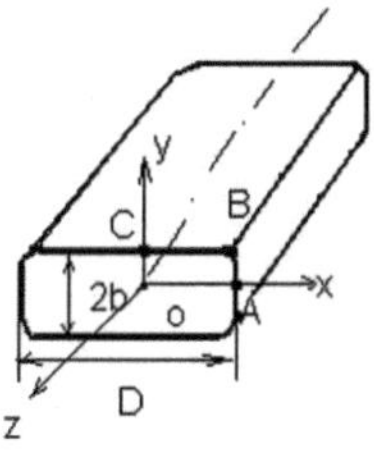

Cross-section form of fuel element	Dangerous points	σ_1	σ_2
	A	$\sigma_z = 0.022(0.019)\frac{\alpha E q_V D^2}{(1-\nu)\lambda}$	$\sigma_\varphi = 0.015\frac{\alpha E q_V D^2}{(1-\nu)\lambda}$
	B	$\sigma_z = 0.027(0.022)\frac{\alpha E q_V D^2}{(1-\nu)\lambda}$	0
	C	$\sigma_x = 0.026\frac{\alpha E q_V D^2}{(1-\nu)\lambda}$	$\sigma_z = 0.011(0.014)\frac{\alpha E q_V D^2}{(1-\nu)\lambda}$
	O	$\sigma_x + \sigma_y = -0.014\frac{\alpha E q_V D^2}{(1-\nu)\lambda}$	

B and C at the same thermal condition are lower than for the round rod of the same diameter. Really, for the round cylinder coefficient at a complex $\alpha E q_v R_2/(\lambda(1-\nu)\sigma_{zz})$ is equal to 0.125, whereas σ_{zz} for the double-blade rod in a point B is equal to 0.108 or 0.088, depending on criterion Bi, that is less accordingly on 13.6 or 29.6 % than for the round cylinder.

For a one-dimensional problem connection; between deformation and stress with account of an elastic deformation and a creep strain becomes: $d\varepsilon_x/dt = B(t)\,\sigma_{xm} + (1/E)d\sigma_x/dt$, where E—Young modulus.

The nonlinear thermal creep problem generally expressed by a numerical method. Calculation is made for the hollow ZrC cylinder which is heated up at a regime of linear increment of a heat release q_v for 3 s to 20 W/mm^3 with continuation of heating during 10 s at this constant power (Fig. 2.6). The properties data: heat conductivity, a coefficient of linear expansion, a modulus, and a speed of creep for calculation are given in Chap. 5.

At the initial stage of the HRA development, different types of FEs, in particular, spherical ones were considered [2, 3]. Spherical HRE systems have a high hydraulic resistance to the cooling gas flow, and therefore rod HREs were preferred. At the final stage, a double-blade twisted fuel element has been chosen from alternative assemblage of cylindrical elements with three and four blades (Fig. 2.7a–c). For decrease of temperature stresses the fuel elements spliced of two, three, and more carbide wires were also offered (Fig. 2.7d–e). The special place occupies a ball fuel element having minimum temperature stress, as it has no ribs, a covering on it are kept better then on other fuel elements designs. Cylindrical fuel element with temperature stresses takes the second place after the ball. It's this property also led to the idea of creation "bladed elements".

Design fuel elements sampling can be made by many criteria. For example, on the criteria of the least hydraulic resistance at the set factor of a convective heat exchange, on the greatest temperature of gas heating at the set maximum fuel element temperature, and also by the technological reasons.

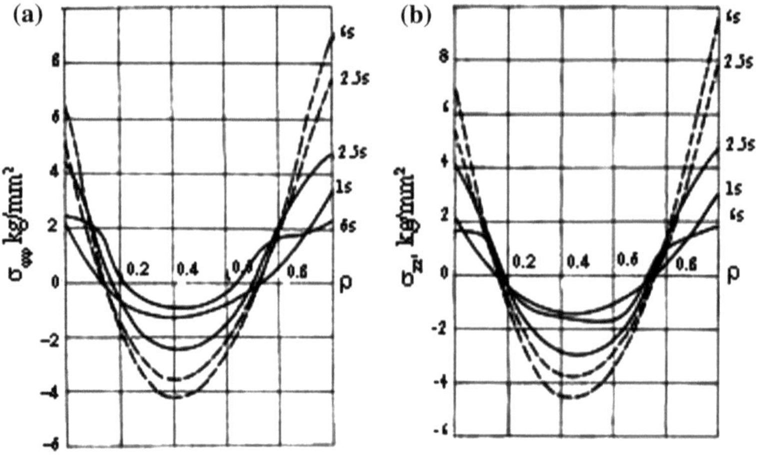

Fig. 2.6 Distribution of the peripheral (**a**) and axils stress (**b**) on a wall thickness of the hollow cylinder. *Dashed lines* a thermal elastic stresses, *full lines* stresses taking into account the relaxation [5]

Fig. 2.7 The sections of various fuel elements types

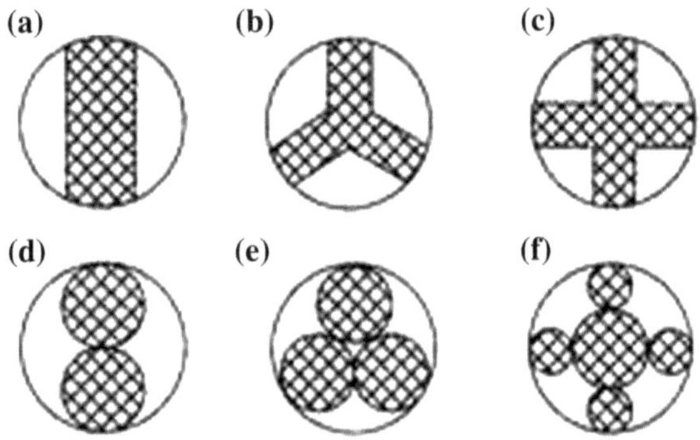

Fuel elements, braded from two wires, with the same diameter, as the double-blade rod possess small advantage on temperature, and more heat release. However, application of the wire fuels was prevented by a high probability of their destruction already in the course of assembly operations owing to enough small strength of the seals between separate wires. As a result there was a danger of emersion of fragments of separate wires. Therefore, this modification was shut down and the double-blade twisted fuel elements was chosen with the worst thermal characteristics, but with the best fracture character not forming small fragments, in comparison with four blade fuel destruction [3].

The heat-transfer agent stream in the HGA with the radial current was formed by distributing and modular collecting channels (Fig. 2.8).

The speed of heat-transfer agent in the radial direction of an order smaller value allows forming a heat reacting surface from the spherical elements of a submillimetric

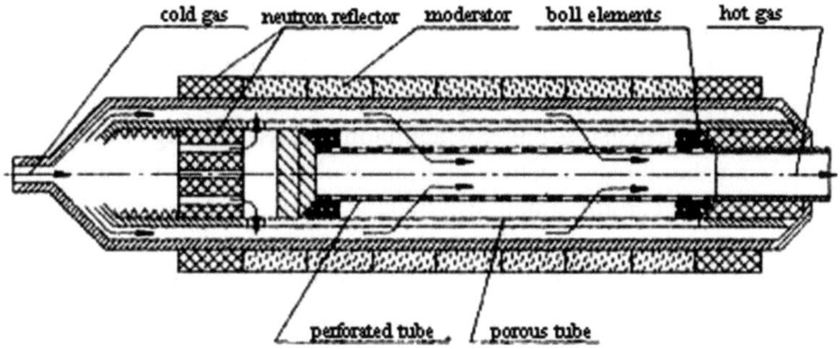

Fig. 2.8 The HGA with a radial working medium flow

or millimetric range. Such ball elements possess the maximum thermal strength resistance. Their dimensions provide more developed surface heat removal and they are perspective for installations with the big thrust of 1,000 kN. Problems of the thermal insulation of HGA casing disappear, since due to the radial hydrogen stream conditions realizing a wall cooling (by blowing of a refrigerating medium through a porous wall). The principal cause because of the radial circuit design of a working medium current yet had no further development, connected at a collecting irregularity of spheres placement, and a problem of flow maintenance of a given working medium along the HGA and absence of a reliable hot wall design of a modular collecting channel.

In the late 1970s and early 1980s RITP and RDIET have commenced an intensive work on developing a multimodal system [2, 4] capable of producing both jet thrust and electricity to power life support systems of the spacecraft. Besides the main nuclear propulsion mode, the NRE was to be operating at two generation modes: low-power mode for prolonged operation (several years), and high-power mode for 1/2 of the specified service life in the propulsion mode. The high power mode (HPM) presented no particular problems for the reactor. At low power mode (LPM) the heat transfer agent circulates only outside the NFA casing, while the heat from fuel rods is transferred to the casing by radiation through the thermal insulation. Such mode differs significantly from the propulsion mode, the former involving considerable temperature gradient across the NFA radius and uranium burnout (min. 3–5 %). Therefore, the applicability of the structural NFA parts and fuel rods under these conditions demands further research. First of all, the design and processing technology of the fuel rods should guarantee retention of fission products inside the rod for several years at temperatures of 2,000 K under high vacuum or in H-containing working medium pressures of 0.1–0.2 bars.

Started almost five decades ago, the program for development of nuclear rocket engine (NRE) originally based on the political aims and priorities of conducting Cold War between the USSR and the US were suspended in the early 1990s due to the USSR having stopped funding of these works.

A heat-insulating packet (HIP) protects the housing from the thermal action of the working substance. Its constructive feature is a multilayer sectional packet structure that minimizes the possibility of penetrating cracks (to the housing) and allows varying the HIP material composition over both its length and its thickness. The outer casings of the HIP made of pyrolytic graphite provide, along with heat insulation, a 'soft' contact with the housing, thereby facilitating the assembling of the construction and minimizing the abrasive action of heat insulation on the housing. The inner casings are thin-wall carbide-graphite cylinders. In the low-temperature region, they are made of zirconium carbide, and in the high-temperature region, they are based on niobium carbide. These casings serve as the supporting frame of the HRA preventing the entry of fragments of heat-insulation elements into the channel of the heating sections. The cases ensure the assembling of HSs and their mounting into a heat-insulating packet and reduce the erosion and chemical action of the working substance flow on the heat insulation. Casings made of low-density pyrographite and porous zirconium and niobium carbides are placed between pyrographite casings and cases. Casings made of low-density pyrographite are located in the low-temperature region ($T = 1{,}500$–$2{,}000$ K). At higher temperatures, casings made of the so-called foliation consisting of carbide layers in a graphite matrix were arranged in the first version of the construction. In the second version, they were replaced by casings made of porous zirconium and niobium carbides.

For the thermal flow density in the cooling channel up to $q_s \ll (2{\div}2.5)\,\mathrm{MW/m^2}$ and the maximum temperature on the surface of the inner case of the packet 3,000 K, the thermal insulation should ensure the temperature on the metal housing of the channel not more than 760 K, which means that the effective heat conduction of the packet should not exceed $3\,\mathrm{W/m\,K^{-1}}$ at $T = 1{,}500$ K.

The temperature distribution over the heat-insulating packet thickness is determined by four heat transfer mechanisms in the gaps: molecular, convection, radiation, and contact resistance. The heat transfer depends on the gas composition and pressure in the gaps, the gap width, the wall temperature, and the gap eccentricity between the walls [6]. Estimates made for two versions of the HIP design (Fig. 2.9) show that the gap between the walls of the casings has the strongest effect on the heat transfer.

A bearing nozzle unit (BNU) supports the HSs and partially supports the HIP. All the axial stresses produced by the pressure drop are transferred through this unit along the hot HRA channel to the housing. In addition, the BNU provides the ejection of the working medium with specified parameters. To minimize the effect of possible cracking, the BNU consists of sections. It contains a bearing grid (BG), a bearing socket (BS), and a nozzle unit. Bearing grids are in the form of a 'sintered' unit of four-blade rods made of solid ZrC and NbC carbide solutions of equimolar composition. To increase the bearing area and provide a cylindrical surface, segment facings are attached to the side surfaces of the rod unit. The bearing socket consists of three successively arranged inserts made of carbide-graphite with a carbide protective covering. The nozzle is made of a set of conical carbide-graphite inserts.

The input unit is intended to produce a uniform gas velocity field at the input to the HSs, face screening of the neutron flux, compensation for thermal expansions of HSs, HIP, and BNU, and tracing of pulsed tubes and thermocouples used for measuring the working substance parameters. The input unit contains springs for compensating temperature expansions, a pyrographite casing, a beryllium cup serving as a face

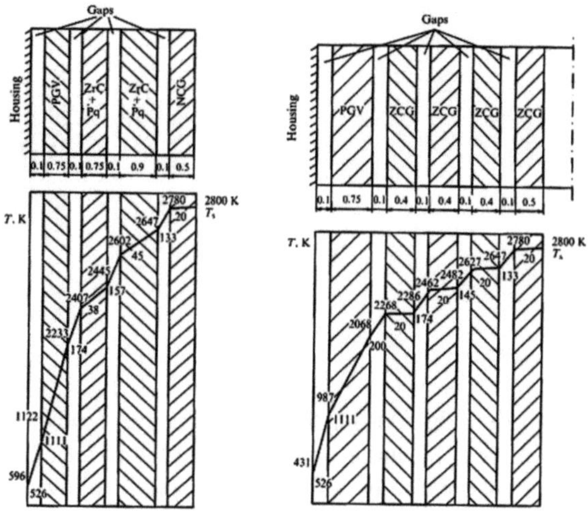

Fig. 2.9 Temperature field in the hydrogen atmosphere in a four-layer packet with two casings made of ZrC + Pq flaking (**a**) and in a five-layer HIP based on three ZrC + C casings (**b**) for the thermal flow through the packet $2 \cdot 10^6\,\mathrm{W\,m^2}$ [7] (*PGV* pyrographite type, *NCG* niobium-based carbographite, *ZCG* zirconium-based carbon graphite)

deflector, a gas inlet, and an input grid consisting of a high-pass grid and four grid rows.

The force elements of the HRA operating at high pressure drops are made of hydrogen-compatible materials having a high specific strength, high radiation resistance, and low hydrogen embrittlement, especially in the soldering and welding joints. The metal housing of the HRA in the IVG-1 reactor in the active core region was made of an AMG-5 aluminum alloy and of 18-10 steel in other regions. The HRA housing in the IR-100 reactor, which should be at the external pressure of the order of 10 MPa during flight, has two layers. The inner layer consisting of beryllium inserts provided the housing stability, while the outer layer, a thin steel jacket, provided the HRA sealing.

Among the zirconium hydride moderators developed and proposed so far, the simplest moderator contains a vertical set of thirteen perforated zirconium hydride discs that are closely adjacent to each other and have the diameter equal to that of the reactor core and thickness 50 mm each. The discs have 37 holes 41 mm in diameter for the HRA and 372 holes 3 mm in diameter for the flow of cooling hydrogen, which provides the required temperature field in the discs. This construction offers a simple solution to the problem of profiling the cooling system of the moderator with the energy release up to $1\,\mathrm{MW{\cdot}cm^{-3}}$ (Fig. 2.10).

The most thermal stressed part of a moderator disk is the average cross-section. The maximum temperature drop at design reactor condition makes ~200 K. The stress-deformed condition of moderating material is defined at two-dimensional problem of thermal elasticity for a case of plane stress deformation. The maximum stress on a surface of cooling channels in diameter of 3 mm with surface temperature Ts = 475–513 K attains magnitude $\sigma_\varphi = 80.5\,\mathrm{MPa}$ [8]. This figure does not provide normal working capacity of moderator on a design condition. Therefore, other alternative of the moderator block with the reduced stresses has been offered.

Fig. 2.10 Neutron hydride moderator block

Besides a disc design, the circuit scheme of rod moderator from zirconium hydride was offered [2], more optimum from the point of view of thermal strength. Such moderator consisting of rods circular profile allows:

- to organize a profiling of an active zone of moderator and to increase efficiency of the reactor;
- to raise thermal strength resistance of the elements making a zone of moderating material at the expense of its size decrease;
- regular distribute energy of dynamic loads in the volume of active zone.

The nuclear reactor deflector block presents itself the hollow cylinder consisting of twelve sectors, each of which contains a compound drum with controlling rods. A rejecter material is a beryllium.

The stationary temperature condition in the deflector block is attained after $\tau = 24\,\mathrm{s}$. The maximum stress occurs at the moment $\tau = 5\,\mathrm{s}$ and exceeds stationary stress on the average more, than in 3 times; their values at $\tau = 5\,\mathrm{s}$, $\sigma_{\theta\mathrm{Max}} = 50\,\mathrm{MPa}$, and $\sigma_{\theta\mathrm{min}} = -120\,\mathrm{MPa}$.

The safety factor on ultimate strength members of the deflector makes 3 that formally meet demands of normative documents.

The basic parameters of the NRE are the temperature, the neutron flux, and the average level of pressure and pressure pulsations of the working substance in different regions of the HRA. A measuring system in the ground IVG prototype [9] provides a reliable control of the working process in the HRA and ensures autonomous emergency protection of the mount over the HRA parameters during tests. In each HRA, two measurements of the gas temperature were provided at the end of the third HS, one measurement of the gas pressure behind the BG and two measurements of the housing temperature in the middle of the HS. The working medium temperature in one cross-section of the HS in the HRA is controlled with two to five zone thermoelectric converters (TECs). Zone TECs made of a tungsten-rhenium alloy are used to measure the inhomogeneous temperature distribution in the radial direction. In general, inaccuracy of measurements is a function of several factors, such as thermal-physical properties of a junction, neutron fluence, temperature, velocity, and pressure of coolant flow. Complex of metrological studies [10–12] that were carried

Fig. 2.11 Input channels of a working medium and communication systems for measurement of pressure and neutron flux at lower part of IVG-1

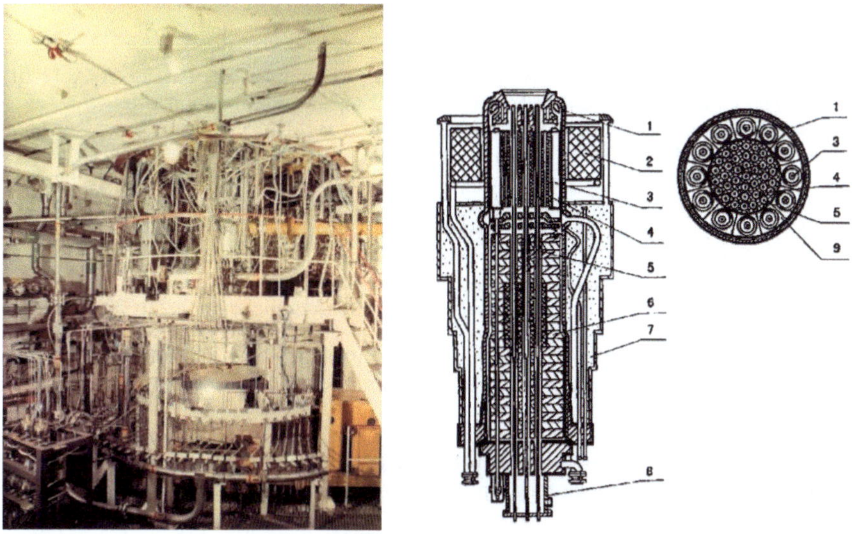

Fig. 2.12 Test reactor RA capacity 0.5 MW: *1* the case in diameter of 586 mm and height 700 mm; *2* additional graphite deflector; *3* regulating drum; *4* moderator; *5* ampoules with fuel; *6* technological console; *7* adjusting mechanism; *8* casing; *9* a deflector

out made it possible to manufacture several thermocouple devices ensuring control of a thermal mode of the fuel assembly testing (Fig. 2.11).

Inaccuracy of measurements of console type thermoelectric transducer is 2.5 %, while antenna type thermoelectric transducer has minimal inaccuracy of measurements which is not higher than 1.9 %; at the same time, durability of the latter device during testing is several seconds only. Small-sized thermal-electric neutron detectors (TEND) with a diameter not more than 2 mm have been successfully used for detecting profiles of the thermal neutron flux density over the height and in different sections of the active zone of the NRE. TENDs do not require the external power supply, and they are not sensitive to a value of isolation resistance under conditions of reactor radiation. Endurance radiation tests confirmed long-term operability of TNDs up to the thermal neutron fluence of 2×10^{21} cm^{-2}. Pressure and temperature of hydrogen at the FA outlet and at the jet inlet define major parameters of the engine, the thrust and specific impulse.

The working capacity of fuel elements for the multimode NRE at low power regimes (LPR) was investigated from the beginning of 1987 [9] in the ampoule design (RA) which was reconstructed from reactor IRGIT No 3, capable to work continuously for a long time (months) for development of various fuel element geometry and compositions (Fig. 2.12).

References

1. Demjanko, J. G., Konuhov, G. A., Koroteev, A. S., Kuzmin, E. P., & Pavelev, A. A. (2001). *Nuclear rocket engines* (p. 413). Moscow: Open Company “Norm-inform”.
2. Nuclear Technology Engineering Industry, Moscow. vol. IV-25, under Adamovs’s edition. Engineering industry., book 1. 953 p. book 2. 943 p.
3. Daragan, I. D., D’jakov, E. K., Fedik, I. I. et al. (2003). Fuel element assemblages of the space nuclear power propulsion systems (vol. IV-25). Moscow: Nuclear Technology Engineering Industry. under Adamov’s edition. Engineering industry, book 2.
4. Ponomarev-Stepnoy, N. N. (1993). Creation history of NRER in the USSR (pp. 3–18). *Third Branch Conferences “Nuclear Power in Space”.*
5. Vlasov, N. M., & Fedik, I. I. (2001). *Fuel elements of nuclear rocket engines* (p. 207). Moscow: Tsniatominform.
6. Korolev, L. A., Lanin, A. G., Morozov A. V. et al. (1978). Research of overall performance of heat-shielding materials and packages of the NRER (vol. 116000, p. 105). Podolsk: Scientific research institute “LUCH”.
7. Zelenskij, D. I., Pivovarov, O. C., Tuhvatulin, S. T. et al. (1999). Experience generalization of reactor working off of the rod carbide fuels on a complex stand “Baikal-1” and the development of applied production engineering (pp. 49–60). *The Fifth International Conference “Nuclear Power Space”, Podolsk.*
8. Golba, A. V., Misevich, J. M., Rachuk, V. S. et al. (2005). Strength problem of design elements of nuclear rocket engines (pp. 405–411). *The International Conference “Nuclear Power in Space”, Podolsk.*
9. Zaharkin, V. I., Ionkin, V. A., Konovalov, et al. (1993). Working out of NRER on the basis of the reactor of minimum sizes IRGIT. Power tests of a developmental type of a nuclear reactor. *Third Branch Conference “Nuclear Power in Space”.*
10. Deniskin V. P., Nalivaev V. I., & Olejnikov P. P. (1999). Problem of metrological maintenance of stand NRER testing (pp. 131–141). *The Fifth International Conference “Nuclear Power in Space”, Podolsk.*
11. Fedik, I. I., Deniskin, V. P., Nalivaev, V. I. et al. (2003). Problem of high-temperature measurements in fuel elements assemblages of NRER (pp. 8–11). The collection of works of Tsniiatominform.
12. Prijmak, S. V., Olejnikov, P. P., & Taubin, M. L. (1989). Features of metrological temperature measurement by thermoelectric temperature transducers in the conditions of intensive reactor irradiation (pp. 327–333). *The Third Branch Conference “Nuclear Power in Space". Atomic Energy 67*(3). 221–222.

Chapter 3
Methods of Modeling Tests

Model tests of the component units of HRAs are, along with calculations, the main tool for the experimental substantiation of the construction at the stage of technical project development. The tests were performed with model samples of component units under laboratory conditions. A specific feature of these studies is the separate influence of different action conditions. The complex action was estimated using the superposition principle by alternating separate tests in stages. This course of action is admissible in the case of the block HRA construction and is caused by the complexity of simulating natural operation conditions. The results obtained on high-temperature testing units allowed optimizing the design and technological solutions used in the HRA development.

At the initial period of NRE development, estimating the efficiency of the construction units of the NRE core was difficult because of the lack of sufficiently complete information on the physical and mechanical properties of materials, especially at high temperatures. It was necessary to verify the correctness of the design and technological solutions in laboratory studies and prereactor tests. To solve these problems, a structure of the experimental equipment was created at the RIHRE, some instruments of the structure being unique. During the development of methods for testing refractory ceramic materials in a broad temperature range from 300 to 3,000 K, a number of difficulties related to the specific properties of these materials were encountered.

The high hardness and high brittleness of carbides, nitrides, and other refractory materials complicating their machining exclude the option of using large samples of a complicated shape because of technological restrictions on their manufacturing. The easy oxidizability of these materials upon heating in air requires performing tests in a vacuum or inert medium and preventing the possible interaction of samples with heater materials.

For these reasons, standard setups for testing metals could not be used and it was necessary to develop or refine a structure of high-temperature setups satisfying these requirements.

A. Lanin, *Nuclear Rocket Engine Reactor*, Springer Series in Materials Science 170,
DOI: 10.1007/978-3-642-32430-7_3, © Springer-Verlag Berlin Heidelberg 2013

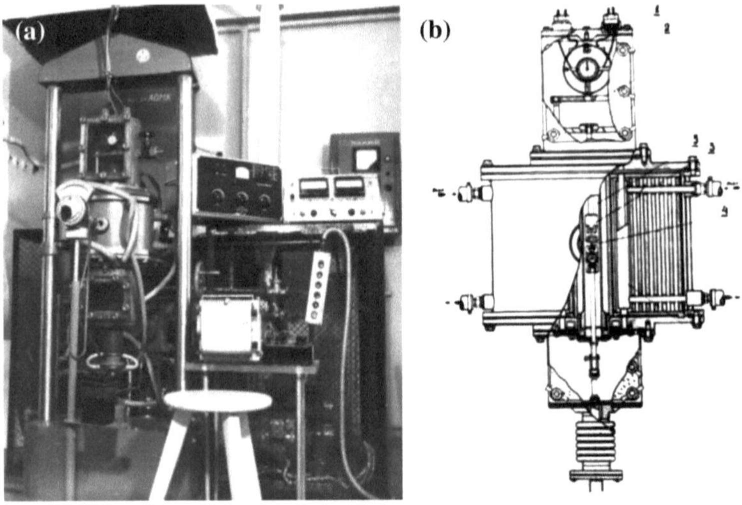

Fig. 3.1 General view of **a** universal high-temperature installation UHTI-1 for mechanical tests up to 3,200 K in argon environment and **b** the scheme of a design of the heating chamber and loading device, (*1* measuring compartment, *2* dynamometer, *3* bars of a sample fastening, *4*, *5* graphite heater)

3.1 Equipment for Measurement of Materials' Strength at High Temperature

Special setups were developed for measuring the strength and plasticity of ceramic materials during bending, compression, tension, and creep and the endurance limit at high temperatures up to 3,000 K in inert gasses or in a vacuum (Fig. 3.1).

When using the dynamometer, the relative measurement error of stresses ranging 0.4–1,000 MN does not exceed 1 %. The measurement error of sample strain, estimated as support displacement difference, was 5 %. This system was used for Russia's first strength test for graphite and ZrC at 3,000 K in 1960. The PRV-203 system built by the Research institute of test machines (RITM, Moscow), originally developed for testing of large samples heated by tungsten heater up to 2,500 K in vacuum (see Fig. 3.2), was subsequently reworked (in particular, the clamping fixture and dynamometer were redesigned) for bending test of smaller samples ($2 \times 2 \times 25$ mm) with stress measurement error of 2 % at the lower load limit of 0.2 MN. Sample strain (with 5 % measurement error) was estimated by support displacement as registered on the device diagram tape at a scale of 1:100.

In 1973, ordered by "Luch" the RITM has developed the electron beam machine EBM-402 for mechanical testing in vacuum at temperatures up to 3,300 K. Sample surfaces were heated by scanning electron beams of two electron guns (see Fig. 3.2a). Samples were subjected to the load of 2–2,000 kgf with adjustable loading speed of 3×10^{-5} to $0.4\,s^{-1}$. Sample strain was measured by extensometer connected to

(a) (b)

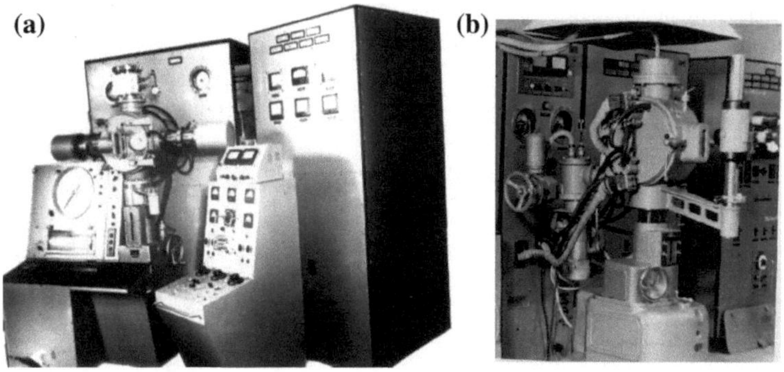

Fig. 3.2 General view of the installation EBM-402 with electron beam heating (**a**) and the machine PRV-203 with the tungsten heater for creep measurement and long bending strength in vacuum or an inert atmosphere (**b**)

variable-induction sensor. Sample temperature was measured by the optical pyrometer, whereas the local temperature pulses in multiple points of the sample were registered by the photovoltaic device. The measurement error of the true sample temperature at 3,000 K was ±86 K.

3.2 Thermal Test Methods

The thermal properties in different temperature ranges were measured by different methods using different setups. Comparative estimates of the thermal diffusivity a, the heat conduction λ, the heat capacity C, and the electric resistance ρ were made by fast transient methods with a typical accuracy of ±10 % in the temperature range from 300 to 1,000 K. The heat conduction of carbide materials at high temperatures (up to 2,800 K) was measured with an accuracy of ±15 % by the axial method by passing current through a rod sample or a fuel element. Simultaneously, the electric resistance was easily measured.

Special technique and equipment were developed for thermal conductivity characterization of the multilayer heat insulated packets (HIPs). The radial heat flux method based on indirect internally applied heating (cylinder technique, see Fig. 3.3) yielded reliable data on heat conductivity λ_{eff} and thermal resistance (R_n) of the HIPs by the radial thermal stream method with indirect heating.

The bearing capacity of thermally loaded constructions was estimated by methods with parameters closest to the operating conditions of devices. The thermal loading of an HRE was simulated by cooling the heated HRE in water, while the thermal loading of cases in the HIP was simulated by heating in melted tin [1]. The thermal strength of materials at temperatures exceeding 0.3–0.5 T_m (where T_m is the melting point), when micro- and macroplasticity can appear, was measured by electron, plasma, and induction heating methods, which require sophisticated equipment. These methods

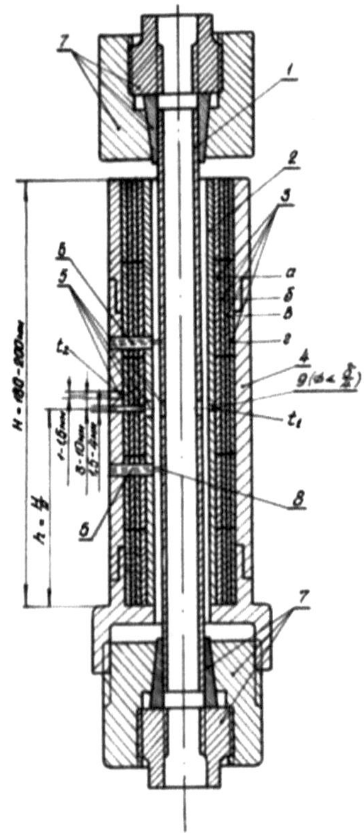

Fig. 3.3 The circuit design of assemblage for measurement of heat conductivity TIP. *1* heater, *2*, *3*, *4* casings: accordingly, internal, testing and outdoor, *5* pyrometric channels, *6*, *8* apertures for potential and potentiometer sounds, *7* a collet system and a crimped lock, *9* the pyrometric channel

were used to vary and control the level and type of temperature fields and the rate and cycle of their variation, thereby permitting quantitative estimates of the thermal strength of materials under conditions close to operating conditions at high temperatures.

The interaction of the HRA units with hydrogen at working temperatures and changes in the composition, structure, and physical–chemical and mechanical properties were investigated using electrical–thermal setups in which the required temperatures could be produced, up to the melting points of materials directly in HREs made of Zr-UC + C and UC-ZrC-NbC. Hydrogen or hydrogen–methane mixture flows passing through working chambers provided the gas rate near the sample surface approximately from 0.1 to 40 ms^{-1} at pressures up to 10 MPa. The hydraulic resistance of the HRA channel in the construction was optimized on hydraulic mounts (Fig. 3.4).

A unit for thermal heat resistance testing by radiation heating under nonstationary conditions includes, as a rule, a hot (a) and a cold chamber (b, c) (Fig. 3.5). The specimen in the hot chamber is heated in an inert medium or in a vacuum to 2,000 °C by means of heater 1, which is a hollow graphite cylinder placed inside an inductor. The heater is shielded from the surroundings by water-cooled double quartz tubes. In

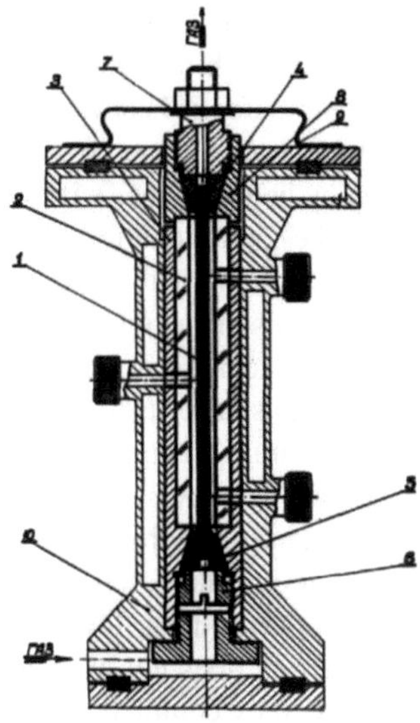

Fig. 3.4 Circuit design of the working chamber of installation B-2. *1* the tested sample (Fuel element), *2* a quartz tube in diameter 10 mm, *3* bearing pipe, *4*, *5* current leads with an air-floating head, *6–9* clutches, the *10* steel case

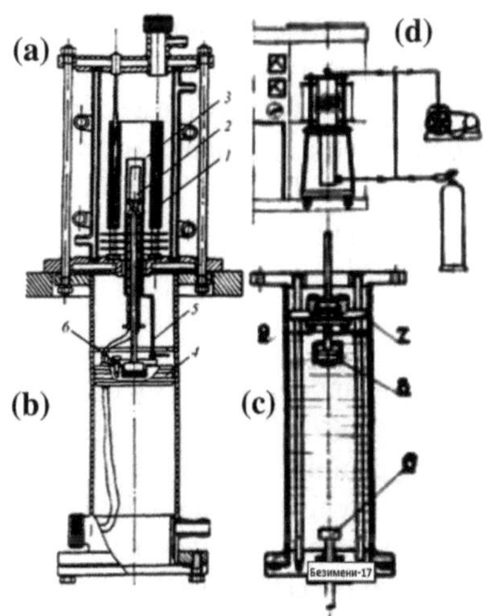

Fig. 3.5 Circuit design of the combined installation RAD for definition of thermal strength resistance by a heating (cooling) method of the sample by radiation or cooling in liquid medium [1]. *1* Heater, *2* specimen, *3*, *5* shield, *4* piston, *6* sound duct

order to create a uniform temperature field along the length of the graphite heater, the heater's edges were protected by molybdenum or graphite shields. The temperature of the heater is monitored by an optical pyrometer. After the heater has reached the desired temperature, specimen 2 (protected on the side face by shield 3) is transferred pneumatically from the cold chamber to the hot chamber as piston 4 is driven by pressure applied from beneath. After the specimen has been positioned in the middle of the heater, shield 5 is released and allowed to drop by gravity, and the side face of the specimen is exposed to the heater. The instant at which the specimen fails is recorded by a piezoelectric sensor connected to the specimen through a sound duct 6; the signal from the sensor is transmitted through an amplifier to a recording oscillograph which is switched on automatically after shield release.

The transfer of heat from the heater at temperature T_h to the specimen's surface at temperature T_s can be treated in terms of constant heat flow $q = const$ with an accuracy better than 5 % if the surface temperature is not in excess of 0.475 T_h [1]; this makes determination of the thermal stress resistance a much simpler procedure.

The maximum relative error in measuring the temperature does not exceed ±0.7 %; this error is associated with the thermocouple calibration error (±0.3 %), measuring device error (±0.05 %), and delay effect error (±0.3 %) which is caused by the heat resistance at the thermocouple—specimen contact through an In–Ga eutectic at temperature change rates of <100 K/s. As the heat flow for the system "heater - specimen" is known, the integral mean temperature drop to failure is determined by merely measuring the heating time to failure τ_f.

The total relative error in the thermal stress resistance composed of the time-to-failure error, heat flow error, temperature extrapolation error, and thermal parameter errors does not exceed 10 % with a probability of 0.68. As a rule, failure is manifested by complete disintegration of the specimen.

It should be noted that the cylinders and disks used as test specimens are more convenient than hollow cylinders or rings. However, in both cases heat resistance of the material in terms of mean integrated temperature ΔT_m does not exceed 100 °C, and the temperature at failure can be varied only within 300–700 °C.

Test process consists in a definition of the temperature pattern at the body fracture moment and a calculation of fracture mean integrated temperature drop in a body $\Delta T_m = T_m - T$, where T_m—mean integrated temperature on cross-section of a body and T is a point temperature (on a surface or on a body center), where thermal stress are defined

$$\sigma = \alpha E \Delta T_m/(1 - \mu), \tag{3.1}$$

where E—modulus of elasticity, α—a coefficient of linear expansion, μ—a Poisson's ratio.

The complex $(1 - \mu)/\alpha E = \Delta T_m$ is accepted to name as the first criterion of thermal strength resistance R [1].

High-temperature thermal strength resistance under electron beam heating was measured on UTU-1 system (see Fig. 3.6) developed at RITM, Moscow. The said technique involves both sophisticated equipment and an elaborate measurement procedure, but its capabilities if high-speed thermal loading of the sample (up to 1,000 K/s) in a wide temperature range up to 2,800 K [2] led to effective simulation of the operating conditions of some HGA blocs in NREs. The cylindrical samples in the UTU-1 system were heated by four electron beam guns placed in a horizontal plane at 90° in a sealed chamber under vacuum (min. 0.01 Pa). Thermal strength behavior at various temperatures was measured by slow heating of the sample to the preset temperature, with subsequent high-speed heating ramp up to the sample fracture. The fracture criterion used was the abrupt slope change of the temperature curve as registered on the oscilloscope tape. Temperature stress calculated using numerical methods based on the first boundary conditions, with surface temperature

Fig. 3.6 General view of electron beam unit UTI-1

being a given function of time. The general measurement error of the thermal strength resistance was estimated at ±15 %.

3.3 Methods of Structural Researches

Metallographic and X-ray diffraction analysis was widely used in the investigation of materials. The lattice period and phase components in carbide compositions were determined by standard methods using URS-50IM and DRON-3 X-ray setups. The chemical and phase compositions of materials were studied by the known instrumental methods of chemical, physicochemical, emission spectroscopy, atomic spectroscopy, and X-ray diffraction analysis. In addition, newly developed methods of porometry, radiometry, mass spectroscopy, and neutron-activation analysis were used.

References

1. Lanin, A. G., & Fedik, I. I. (2008). *Thermal stress resistance of materials* (p. 239). Heidelberg: Springer.
2. Popov, V. П., Lanin, A. G., & Bochkov, N. A. (1984). *Test method of thermal stability of samples from brittle electric-conductive materials with use electron beam heating*. New York: Plenum Publishing Corporation (Strength of materials).

Chapter 4
Materials of the Reactor Core

By the beginning of work in the 1960s, information on the properties and manufacturing technology of materials for the NRE core (based on zirconium, niobium, uranium carbides, and zirconium hydride) was absent or inconsistent. It was known that unlike mono compound of uranium with a low melting point (2,500 K), a fuel based on solid solutions of UC–ZrC and UC–NbC carbides with nearly stochastic composition can provide the heating of hydrogen up to 3,000 K. Therefore, investigations of solid solutions of uranium monocarbide with isomorphous, highly refractory zirconium, niobium, and monocarbides providing high melting points and compatibility of HREs with heat carriers became the most important material technology direction. The prospects of the development of UC–ZrC–ZrN fuel were also outlined. The manufacturing technology of these refractory materials was based on powder metallurgy methods.

Refractory compounds are characterized by high hardness, elastic modulus, chemical stability, heat resistance and the high brittleness caused by features of interatomic interaction with mixed ionic–covalent type and low dislocation mobility [1–3]. Graphite with low values of hardness, strength and an elastic modulus possess many times higher thermal stress resistance $R \geq 700$ K in comparison with carbide compounds.

The first data on the radiation resistance of HREs (integrity, swelling, and strength) at temperatures from 1,000 to 3,100 K and neutron flux intensities up to 10^{15}–$10^{16}\,\mathrm{cm}^{-2}$ confirmed the expediency of the choice of fuel materials based in solid carbide solutions. It was decided [4, 5] for the first time in the development of a highly reliable construction in machine building to use brittle materials, which required changing construction principles and the established concepts of strength and thermal strength. A new criterion for estimating the bearing ability of thermally loaded products was introduced, which was accepted by the scientific community worldwide [6].

A. Lanin, *Nuclear Rocket Engine Reactor*, Springer Series in Materials Science 170,
DOI: 10.1007/978-3-642-32430-7_4, © Springer-Verlag Berlin Heidelberg 2013

4.1 Thermodynamic and Structural Characteristics of Materials

Thermodynamic studies of the solid solutions of refractory compounds with uranium monocarbide were started at the RIPRA "Luch" when no data on the thermodynamic properties in the homogeneity region at high temperatures were available in the literature. Only basic approaches for estimating thermodynamic properties were known. Structural and fuel materials of the reactor core are based on refractory zirconium and niobium carbides, their solid solutions with uranium, carbide compositions with carbon inclusions [4, 5, 7], and zirconium hydrides for a moderator unit (Table 4.1). These materials belong to the class of so-called interstitial phases [8]. The majority of refractory compounds possess highly symmetric cubic lattice. The majority of monocarbides, mononitrides crystallizes in FCC lattice of NaCl type, nonmetallic atoms located in octahedral positions. Melting point, elastic properties and factors of thermal expansion are structurally tolerant characteristics and, in essence, they depend on bonding energy, energy of a crystal lattice necessary for division into separate ions.

High brittleness of compounds is caused by low dislocation mobility owing to an orientation of bonding, high value of forces Peierls–Nabarro and low rate multiplication of dislocations. Dislocations submit to the same laws in compounds, as in metals. However in polyatomic compounds with ionic–covalent bonding, more complex geometry of dislocations' structure is observed. Starting stresses of dislocations' movement in crystals with high Peierls barrier are on one to two orders above, than in metals. The stresses demanded for the yield beginning, in turn, exceed start stresses on two orders at the moderate temperatures $0.3\,T_{ml}$.

The data on the physical–mechanical properties of many refractory compounds and their compositions (some of them were used in devices for the first time) revealed the possibilities of these materials and influenced the determination of their operating conditions and estimates of their prospects [9–11]. Unlike uranium mono compounds, fuel based on solid solutions of UC–ZrC and UC–NbC carbides with a composition close to the stoichiometric composition provide the heating of hydrogen up to 3,000 K.

4.1.1 Melting Temperature and Evaporation

Materials for NREs should have, along with high melting points, low evaporation rates and should weakly interact with hydrogen. Changes in the material composition caused by evaporation or interaction with hydrogen should not remove the composition from the homogeneity region during a specified operation time resource.

Melting temperature of carbides in the homogeneity range does not change monotonously. Typically, the maximum melting temperature is demonstrated by the non-stoichiometric phases with composition close to C/Me ratio of 0.8. The melting temperature of solid monocarbide solutions is expected to change similarly in the

Table 4.1 Averaged physical characteristics of reactor core materials in the temperature range from 300 to 700 K [14]

Compound formula	Structure type[a]	Density ($g \cdot cm^{-3}$)	Melting points (T_m K)	Linear expansion coefficient ($10^{-6} K^{-1}$)	Heat conduction (Wm^{-1})	Elastic modulus (GPa)	Vickers hardness (GPa)
Fuel materials							
UC	C	12.9	2,500	10.4	19	220	9.0
UN	C	14.4	3,074	9.3	18	265	8
ZrC + 5 %UC	C	6.9	3,380	11.8	30	380	25
ZrC + 5 %UC + C	C	6.6	3,250	11	32	350	20
ZrC + 5 %UC + Nbc	C	7.6	3,520	11	22	320	28
Construction materials							
ZrC	C	6.73	3,690	8.6	30	390	27
ZrC + 5 % C	C	6.5	3,180	5.5	52	230	18
NbC	C	7.8	3,870	7.7	20–30	500	20
ZrC+50 % NbC	C	7.3	3,620	5.9	25	470	28
$ZrH_{1.9}$ ε-phase	T	5.6	2,470[b]	7.0	30	69	0.16
Pyrographite	H	1.7	4,000[c]	8.5	70	48	0.1

[a] *C* cubic structure; *T* tetragonal structure, *H* hexagonal structure [b] Value for the hydrogen pressure 100 MPa [c] Sublimation temperature

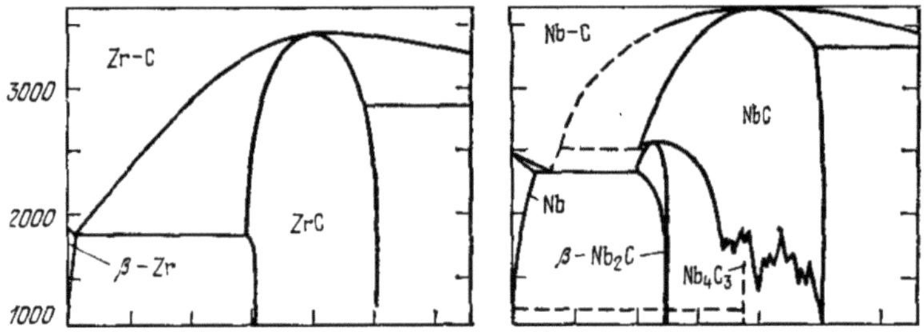

Fig. 4.1 State diagram: Zr-C; Nb-C [2, 3]

Fig. 4.2 The phase diagram of alloys ZrC–NbC–UC, quenched from temperatures 2,100, 1,900, 1,700 K

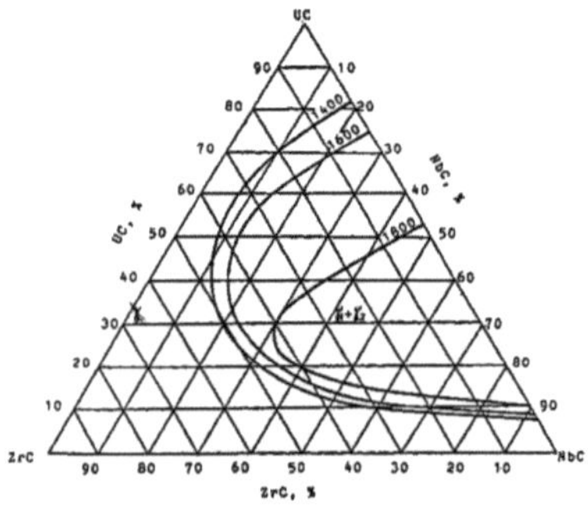

homogeneity range [12]. The melting temperatures of fuel compositions based on ZrC, NbC, and ZrN decrease as the UC content increases (Figs. 4.1 and 4.2). The measured absolute values of melting temperatures for nuclear fuel rods are close to solidus line for the ZrC–UC and ZrC–NbC–UC solid solutions and average around 3,570 and 3,520 K, respectively, being significantly higher than carbon nitride melting temperature. Eutectic compounds from the typically used range of carbide–carbon compounds have the following melting temperatures: ZrC + C: 3,180; NbC + C: 3,580; TaC + C: 3,715 K; the said melting temperatures being lower than that of pure carbides. Evaporation processes have a considerable effect on the performance of the HGA parts and on the behavior of these materials at high temperatures under processes involving mass transfer through vapor phase. In considering evaporation rates of alloys and compounds one should take into account not only the integrated evaporation rates, but also the partial evaporation rates. This is particularly important for such compounds as carbides, nitrides, etc., that demonstrate broad scatter of partial component evaporation rates [13]. The methods for determining the partial thermodynamic functions of two- and three-component interstitial phases, based on statistic and thermodynamic approach are developed in [2]. These methods can determine changes of gas pressure in the homogeneity range of congruently evaporating phase compositions (i.e. the compositions which remain practically unchanged during evaporation) (Table 4.2).

Table 4.2 Superficial V and linear V_1 speeds of evaporation of carbides

Material	V (g/sm^{2}s)		V_1 (sm/s)	
	2,500 K	3,100 K	2,500 K	3,100 K
$ZrC_{0.85}$	$1.21 \cdot 10^{-6}$	$8.37 \cdot 10^{-4}$	$1.824 \cdot 10^{-7}$	$1.257 \cdot 10^{-4}$
$NbC_{0.77}$	$2.99 \cdot 10^{-9}$	$5.03 \cdot 10^{-6}$	$3.83 \cdot 10^{-10}$	$6.5 \cdot 10^{-7}$
$TaC_{0.5}$	$6.15 \cdot 10^{-9}$	$9.83 \cdot 10^{-6}$	$4.16 \cdot 10^{-10}$	$6.64 \cdot 10^{-7}$
UC-ZrC	$3 \cdot 10^{-6}$(2,270 K)	$1.9 \cdot 10^{-4}$		
UC-UN	$8.2 \cdot 10^{-7}$(2,800 K)	$1.8 \cdot 10^{-4}$		

Predicted values for NbC are in good agreement with the experimental data; for ZrC one has to take into account the concentration dependence of the vacancy formation energy [2, 13].

As uranium has higher thermodynamic activity in the $U_{1-y}Me_yC$ solid carbide solutions, compared to that of Zr, Nb, or Ta, and as there are no congruently evaporating compositions in ternary systems, consequently, the uranium loss is expected to prevail in evaporating from an open surface, with corresponding surface enrichment in Zr, Nb, or Ta atoms. Simultaneously, the condensate deposited on a cold wall should have higher U/Me ratio than the one in the initial carbide.

Noncongruent nature of $U_{1-y}Me_yC$ solid solution evaporation makes rather problematic their prolonged use at high temperatures in vacuum. Special protection measures are required. In vacuum metallic component of NbC_x evaporates much slower than that of zirconium carbide and, despite larger cross-section values of thermal neutrons absorption in Nb (as compared to Zr), NbC may be effectively used in carbide–graphite compositions, in binary or ternary carbide solutions, and most particularly, as protective coatings on graphite parts.

4.1.2 Diffusion Characteristics

The statistical thermodynamic theory of three-component interstitial phases developed at the institute and the established properties of diffusion-controlled processes [12] such as sintering, nitration, carbonizing treatment, and oxidation facilitated the optimization of technologies used in investigations. Diffusion processes have a significant impact on the HGA materials' interaction with hydrogen, and on consolidation of carbide powders [3]. Experimental diffusion data were used to estimate the load-bearing capacity of fuel rods at high temperatures and the healing of surface defects. Kinetics of diffusion phenomena in the majority of cases is controlled by a number of elementary processes, whose individual contribution is rather difficult to differentiate.

Thus, densification rate during sintering and creep rate are controlled by volume, boundary, and surface diffusion [12], and by dislocation sliding. Two-component or multicomponent nature of interstitial phases generally complicates the analysis of diffusion-controlled processes, as the contribution of metallic and nonmetallic atoms in various temperature ranges may vary. Lastly, the comparison of the experimental and the theoretical data on diffusion-controlled processes requires clarification of the concept of an effective self-diffusion factor D^{ef} which is usually introduced for the multicomponent bodies [2].

For a single-component body D is a factor of a self-diffusion; for two- or multi-component bodies one should use substitute D^{ef} into the respective equations:

$$D^{ef} = (D^*{}_A D^*{}_B / C_B D^*{}_A + C_A D^*{}_B) \cdot g,$$

where g is the thermodynamic factor, C_i is concentration of a specific component, D_i is partial (chemical) self-diffusion factors that are typically defined through experiments with radioactive isotopes. If the diffusion mobility of one component in the alloy or compound is non-negligible (for example, diffusion mobility of interstitial elements), and $D^*{}_B > D^*{}_A D$, then the equation for the D^{ef} may be simplified as follows:

$$D^{ef} \sim (1/C_A) \cdot D^*{}_A g$$

In this case the rate of diffusion-controlled processes is defined by diffusion rate of the slowest diffusing element [2]. The developed statistic and thermodynamic theory of three-component interstitial phases, and our research of diffusion-controlled processes (creep, sintering, nitriding, carburization, and oxidation) make it possible to formulate evaluation recommendations concerning composition, processing, and operating conditions of NRE fuel elements (Table 4.3).

4.2 Processing Technology of the Structural Ceramic Materials

Furnaces developed and constructed at the institute for sintering and consolidating powder samples and deposition of coatings on products at high temperatures (up to 2,500–2,700 K), which were unique for that time, were used to prepare samples required for studies and, later, components of the NRE core.

The general ceramics processing technology involves the following steps [14]:

1. Synthesis of powders with a given chemical composition, phase composition and grain size distribution;
2. Preparation of powders for forming;
3. Green part forming;
4. High-temperature consolidation;
5. Process quality.

Sintering is a crucial stage for structure evolution. Sintering is a thermally activated process (spontaneous or involving the application of external force) which features transformation of a system of contacting solids or porous media into thermodynamically more stable state through minimization of the free surface area. The driving

Table 4.3 Diffusion parameters of the carbide compounds

Composition	Diffusion element	Temperature range (K)	D_o (sm^2/s)	Q ($kcal/g_{form}$)	D (sm^2/s) at T, K		
					2,500	2,800	3,100
$ZrC_{0.97}$	C^{14}	2,500–3,100	14.1	1089.9 ± 6.1	$4.26{\cdot}10^{-9}$	$4.35{\cdot}10^{-8}$	$2.878{\cdot}10^{-7}$
	Zr^{95}	2,500–3,100	1,030	172 ± 10.7	$9.93{\cdot}10^{-13}$	$3.83{\cdot}10^{-11}$	$7.62{\cdot}10^{-10}$
$ZrC_{0.70}$	C^{14}	2,500–3,100	2.8	164 ± 5.1	$1.40{\cdot}10^{-10}$	$5.09{\cdot}10^{-9}$	$8.095{\cdot}10^{-8}$
	Zr^{95}	2,500–3,100	1,030	172 ± 10.7	$9.93{\cdot}10^{-13}$	$3.83{\cdot}10^{-11}$	$7.62{\cdot}10^{-10}$
$NbC_{0.97}$	C^{14}	2,500–3,100	0.11	94 ± 1.8	$6.62{\cdot}10^{-10}$	$5.25{\cdot}10^{-9}$	$3.24{\cdot}10^{-8}$
	Nb95	2,500–3,100	1.47	127 ± 7	$8.3{\cdot}10^{-13}$	$1.32{\cdot}10^{-11}$	$1.23{\cdot}10^{-10}$
$NbC_{0.79}$	C^{14}	2,500–3,100	0.11	100 ± 1.7	$2.61{\cdot}10^{-9}$	$2.27{\cdot}10^{-8}$	$1.31{\cdot}10^{-7}$
	Nb95	2,500–3,100	0.11	127 ± 7	$8.73{\cdot}10^{-13}$	$1.32{\cdot}10^{-11}$	$1.23{\cdot}10^{-10}$
$(Zr_{0.48}Nb_{0.52})C_{0.9}$	C^{14}	2,500–3,100	2.28	96.5 ± 2.5	$8.26{\cdot}10^{-9}$	$6.72{\cdot}10^{-8}$	$3.56{\cdot}10^{-7}$
	Nb^{95}	2,500–3,100	51.0	153.0 ± 9.8	$2.12{\cdot}10^{-12}$	$5.71{\cdot}10^{-11}$	$8.26{\cdot}10^{-10}$
$(Zr_{0.48}Nb_{0.52})C_{0.8}$	C^{14}	2,500–3,100	0.84	100.5 ± 3.4	$1.36{\cdot}10^{-9}$	$1.20{\cdot}10^{-8}$	$6.91{\cdot}10^{-8}$
	Nb^{95}	2,500–3,100	51.0	153.0 ± 9.8	$2.12{\cdot}10^{-12}$	$5.71{\cdot}10^{-11}$	$8.26{\cdot}10^{-10}$
$TaC_{0.98}$	C^{14}	2,650–3,200	3.9	118.7	$1.66{\cdot}10^{-10}$	$2.58{\cdot}10^{-9}$	$1.66{\cdot}10^{-8}$

Fig. 4.3 Vacuum high-temperature electric furnace $T_{max} = 2,750\,°C$

Fig. 4.4 Installation for hot pressing (IHP-4) with pressing force up to 200 tons and working space 900×1800 mm

force for this transformation is the excess free energy. Sintering is evident both as a dimensional change (mostly as shrinkage), and as significant change in structure and properties, both approaching these of compact materials. At the RIHRE "Luch", there were developed special electrothermal vacuum systems for synthesis, sintering, and hot pressing of refractory carbides, and systems for synthesizing fine carbide powders (Figs. 4.3 and 4.4).

Details of kinetics of density, structure, and properties evolution during zirconium carbide sintering are described [15, 16]. ZrC powders containing 87.8 % Zr, 11.4 % of C_{total}, 0.15 % of C_{free}, 0.018 % N, 0.62 % O were synthesized by carbon-thermal reduction at $T = 2,100\,°C$. The synthesized powders were ball-milled in a vessel lined with refractory ZrC plates. Average particle diameter of powders was 4.5–5 μ, with BET surface area of 4.35 m^2/g. Green parts were formed by thermoplastic extrusion, dewaxed in a vacuum furnace at final temperature of 350 °C with heating rate of 10 K/h, and dwell time of 2 h. Final sintering was carried out in the graphite

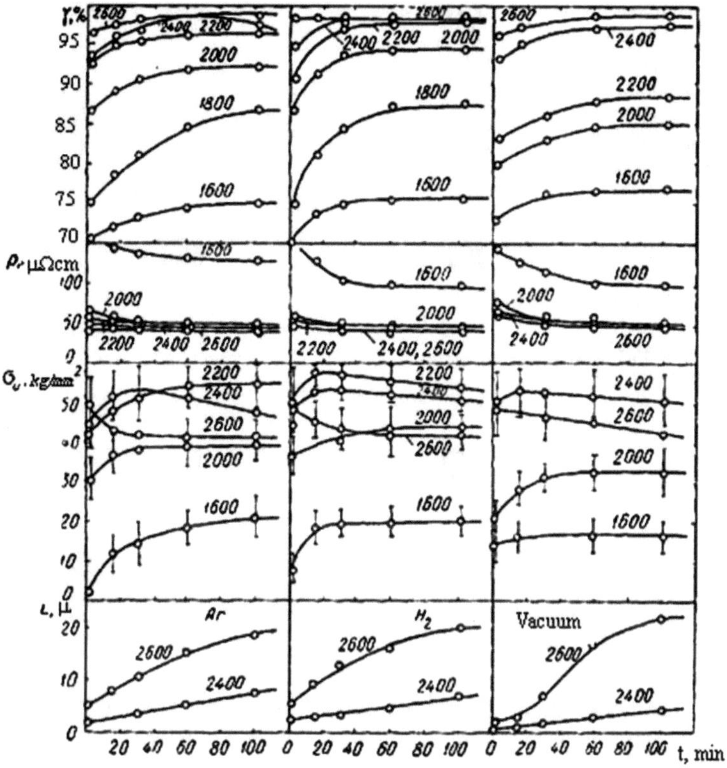

Fig. 4.5 Kinetics of a relative density change (γ), electrical resistance (p), bending strength (σ_b), and magnitudes of grain size (L): at sintering of zirconium carbide in an argon, hydrogen and vacuum of 10^{-3} mm Hg (temperature in °C)

heater furnace at 1,400–2,800 °C with dwell time of 0, 15, 30, 60, and 120 min; heating and cooling rates were 800–900 K/h (Fig. 4.5).

High values of γ, σ and electrical conductivity were obtained at 2,200–2,400 °C. Sintering in hydrogen and argon atmospheres yielded similar results, while vacuum sintering yielded slightly poorer properties, which is attributable to greater gas evolution during sintering possibly leading to more porous compact structure. Similar results were observed during sintering of ZrC+NbC and ZrC+NbC+UC systems [11]. The further increase of sintering temperature to 2,600–2,800 °C did not lead to better density or strength of the sintered samples. Poorer sample consolidation could be due to the effect of gases in the closed pores. At the overall porosity of ~6–7 %, the majority of pores are closed which greatly hinders further consolidation.

Higher porosity of samples sintered at 2,800 °C may be attributed both to gas pressure and to coalescence of gas-filled pores. Lesser strength of samples sintered at 2,600–2,800 °C is due not only to poorer density and significant grain coarsening, but also to pore segregation at grain boundaries leading to formation of the

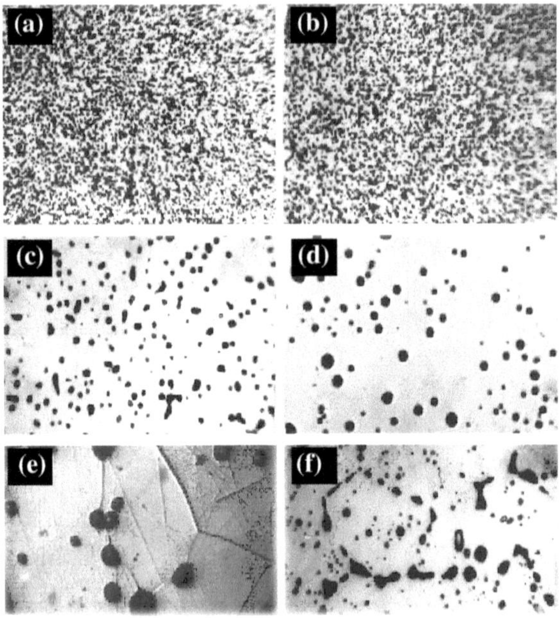

Fig. 4.6 Features of microscopic ZrC structure sintered at different temperatures and various medium [16]. **a** Vacuum, $T_S = 2{,}000\,°C$, $\tau = 15\,min \times 1000$. **b** Vacuum, $T_S = 2{,}200\,°C$, $\tau = 15\,min \times 1000$. **c** Vacuum, $T_S = 2{,}400\,°C$, $\tau = 15\,min \times 1000$. **d** Vacuum, $T_S = 2{,}600\,°C$, $\tau = 15\,min \times 1000$. **e** Hydrogen, $T_S = 2{,}700\,°C$, $\tau = 5\,min \times 1000$. **f** Argon, $T_S = 2{,}700\,°C$, $\tau = 5\,min \times 1000$

highly specific crack patterns (Fig. 4.6). Samples sintered at $T > 2{,}600\,°C$ show predominantly intergranular fracture, while those sintered at lower temperatures show mostly transgranular fracture. The relative density and strength σ of fuel rods based on ZrC+UC and sintered at 2,200 °C in hydrogen reach 99 % and 650 MPa, respectively, while vacuum sintering at the same temperature yields values of 88 % and 360 MPa, respectively [11].

When analyzing the sintering process one should consider porosity features. At relative density nearing 90 % of TD, the residual porosity is nearly always an open one, i.e., pores are connected to a surface. At overall porosity of 7–8 %, all pores became closed. As temperature rises or as sintering time increases, the powder compact becomes denser and the shrinkage increases, possibly reaching 20–25 %. Simultaneously, there is a change in properties: density change could be traced by monitoring the decrease of the electrical resistance.

During sintering in gas atmosphere at 2,000–2,400 °C, and in vacuum at 2,300–2,500 °C densification mainly proceeds by decreasing the closed porosity; further temperature rise to 2,600 °C yields virtually no change in closed porosity. At this stage, pores coalesce and become spherical (see Fig. 4.6c, d). The number of pores decreases, while their size increases. At sintering temperatures above 2,600 °C, the closed porosity and overalls porosity values grow. Sometimes surface cracks appear on samples, leading to decrease of the closed porosity. This particular phenomenon was most frequently observed in samples sintered in argon at temperatures above 2,700 °C.

Bubble-shaped pores at this stage of sintering are pinned by boundaries at which these pores tend to segregate and coalesce (see Fig. 4.6e, f). The shape of the pores changes, vast cavities appear, and part of bubble pores migrates to the sample surface as a result of coalescence of two gas filled pores of equal radius; the volume of a newly created pore may increase (provided that the volume if gas before and after the coalescence remains constant), which leads to the swelling of samples. In ZΓC samples sintered at temperatures above 2,600 °C, the volume increase and density drop were observed. In some samples, this phenomenon was observed already at sintering temperature of 2,550–2600 °C, especially in argon atmosphere. The results of analysis of the furnace chamber atmosphere show that during sintering in Ar at pore closing stage (T at 2,000–2,200 °C) the major impurity is CO which may become trapped in the closed pores of the sample. Moreover, CO also forms during sintering through the reaction of free carbon with oxygen, and during rapid heating at 2,000–2,200 °C this CO may also become trapped in pores.

Samples with the green density ρ of 52–57 % were sintered in vacuum at temperatures of 1,000–2,250 °C, and subsequently in Ar atmosphere at T_{max}= 2,500–2,700 °C. The shrinkage kinetics during sintering in a furnace with the special fixture for controlling the linear dimensions of the sample was constantly measured with an accuracy of $\pm 10\,\mu$.

The sintering behavior of the system "ZrC+ 2÷10 % C (diamond) up to the temperature of 1600–1800 °C was similar to that of the pure ZrC [17]. In the temperature range of 1,650–1750 °C corresponding to the maximum volume change of the diamond (due to its polymorphic transformation), the shrinkage rate dropped nearly to zero. Further temperature increase for this composition leads to significantly accelerated densification. At this stage, the sample shrinkage curve of the ZrC–2.5 % C system is similar to the one of pure ZrC. Changing the diamond particle size from 50 to 1 μ leads to a change in shrinkage curves. The absolute value of the density loss (caused by polymorphic transformation of diamond) decreases. The sintered compositions being studied show predominantly aggregated structure with carbon phase particles located primarily on the grain boundaries of the carbide matrix.

The general regularity of structure formation of ZrC, ZrC–C, NbC–C is the high speed recrystallization at sintering temperature above 2,250 °C: The grain sizes of compositions increase to 10–20 μ with growth of carbon phase after sintering at 2,500 °C. Below this sintering temperature, the grain size of sintered composition are comparable with initial sizes of carbon corpuscles (5 and 2.5 μ).

The pores placed on boundary lines of grains, suppress grain growth at sintering. An average grain size of carbides increases slightly (from 20 to 25 μ) at the sintering temperature interval 2,500–2,700 °C. As a whole the structures of sintered compositions ZrC, NbC with carbon additives (diamond, graphite) is characterized by the uniform distribution of pores in a matrix [18]. Estimations of the sintering energy activations Q of compositions ZrC–C, NbC–C show that Q in the investigated interval of temperatures is changeable [17]. At temperatures 1,300–1,400 °C $Q = 65-80$ kkal/g-form for compositions ZrC–C and NbC–C $Q = 70-75$ kkal/g-forms. At higher sintering temperature Q increases to 115–125 kkal/g-forms for (ZrC–C) and to 120–135 kkal/g-forms for (NbC–C). The received values Q at

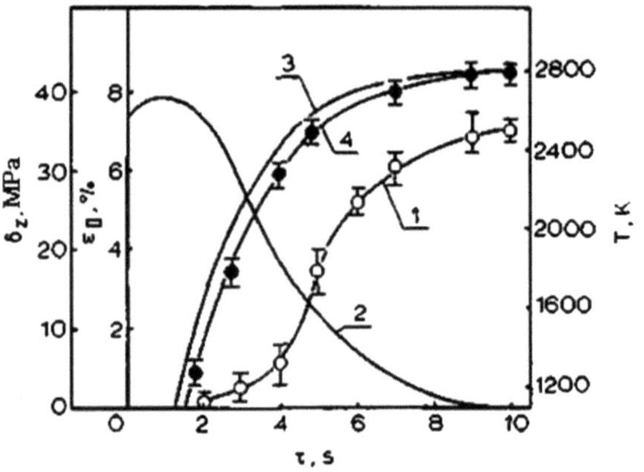

Fig. 4.7 Kinetics of a shrinkage ε_{II} (*1*), a thermal elastic stress (*2*), the calculated (*3*) and measured (*4*) surface temperatures of the pressing $ZrC_{0.95}$ in diameter of 2.4 mm in the course of high-speed heating to maximum temperature 2,800 K [19]

sintering of the compositions ZrC–C, NbC–C proceeding from the parameters of a self-diffusion component in carbides ZrC, NbC [2], testify to predominance of the boundary diffusion contribution in mass transfer at sintering to 1,300–1,400 °C. Role of the volume diffusion controlling the shrinkage at sintering becomes appreciable at temperature growth.

It is known that the temperature pattern formed in a powder pressing at heating–cooling can essentially influence on the character and intensity of sintering. However, the data on an agency of heating and cooling speeds on a process of carbide materials sintering in the literature are restricted by consideration of small speeds of heating–cooling near 500–1,000 K/h. In this connection, an undertaken attempt [19] to use high-speed heating–cooling when temperature-kinetic parameters of heat treatment can become crucial. High temperature heating up to 2800 K is used for the carbide pressings $ZrC_{0.95}$ in diameter of 2.5 mm with a speed 600 K/s allows for increment sharply to 10^3 times speed of a shrinkage and to raise an initial density from 71 to 98 % (Fig. 4.7).

The influence of sintering parameters on the sintered ZrC structure and properties are showed on Figs. 4.8 and 4.9 accordingly. Sintering of ZrC samples by rapid nonisothermal heating takes place in a complicated manner. In the first stage of the sample heating to 1,300 K, for the first 2 s thermal elastic stresses of the order 30–40 MPa (Fig. 4.7) are developed and are capable of creating local plastic deformation of the material on rough surfaces of the particles. The presence of this phenomenon is supported by an experimental study of the dislocation start from the indentor print on the single crystal $ZrC_{0.9}$ performed at a temperature of 290 K and subsequent heating to temperatures 600 and 850 K. The initial stresses σ_s for dislocation movement at these temperatures are at the limit of 5–8 and 2–3 MPa, respectively [20]. It should be noted that local stresses on roughnesses with the size of hundreds of thousands of angstroms can exceed average-mass thermal elastic stresses by one to two orders. And the latter can relax at the very initial stages of sintering.

In the second stage, which occurred with further temperature increase up to 2,000 K, the shrinkage, perhaps, is defined by mutual slip of particles as a whole

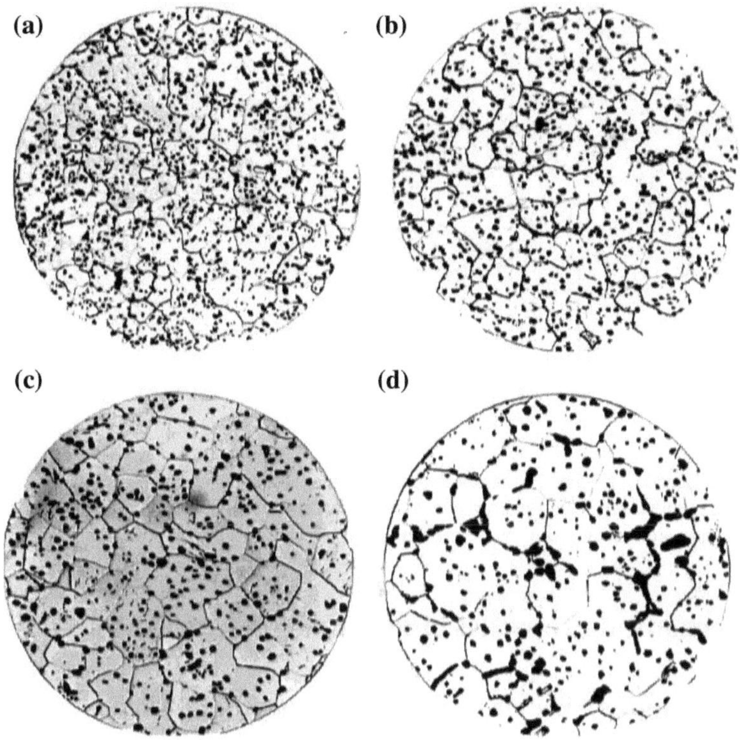

Fig. 4.8 Influence of a maximum heating temperature T at high-speed sintering $\delta T/\delta t = 600$ K/s on the ZrC structure. **a** T = 2,300 °C; 4 μ ×300. **b** T = 2,500 °C; 7 μ ×300. **c** T = 2,700°C; 12 μ ×300. **d** T = 2,900°C; 30 μ ×300

along the boundaries between them. They move into the voids and to the free surface of the sample (Fig. 4.8a–d) according to the mechanism proposed in [19, 21].

After overcoming the resistance of retainers on the definite part of the surface of the particles and their 'activation' the boundary diffusion viscosity becomes low and shrinkage sharply increases. Random boundaries that are not connected with initial boundaries between particles are formed. Shrinkage rate decreases with porosity decrease.

At the third stage of heating to temperatures higher than 2,200 K in the central part of the sample, the final stage, coagulation of pores takes place on far from the free surface, isolated pores appear. Further densification takes place due to decrease of the peripheral porous layer with open pores and a reduction in the number and volume nonisothermal sintering of the samples at mainly of the isolated pores (Fig. 4.8d).

An appropriate choice of the nonisothermal heating parameters can provide a fine-grain regular-pore structure (Fig. 4.8b), with high level of strength at appreciable time decrease of sintering.

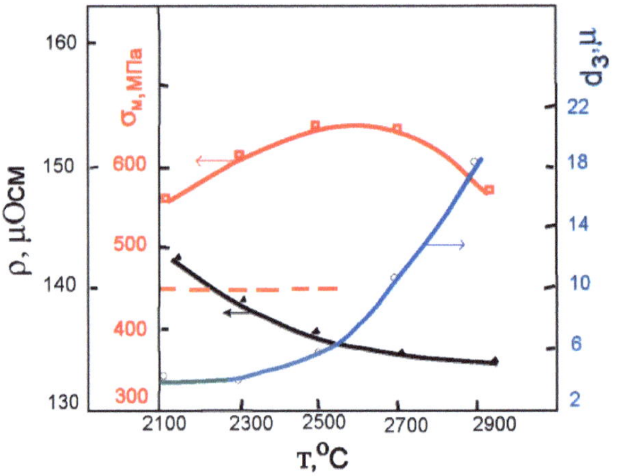

Fig. 4.9 Influence of a heating temperature at high-speed sintering $\delta T/\delta t = 600$ K/s on ZrC properties: grain size d_3, electrical resistance ρ and bending strength σ_b. *Dotted line* is the strength of ZrC sample sintered at ordinary speed after an exposure 1.75 h at 2,500 °C

4.3 Mechanical Properties of Fuel and Structural Materials

4.3.1 Strength of Materials at Different Loading Mode

The change of loading mode from tension to bending and compression for ZrC, NbC, ZrC+NbC at the deformation rate 10^{-3} s^{-1} decreases brittle–ductile transition temperature T_{b-d} and rises the level of ductile deformation. The strength level of ceramics in brittle condition in most cases depends on the stress state considerably so while passing from tension to bending and compression the strength increases in relation 1:(1.5–2.0):(8–10) (Table 4.4) [14]. This phenomenon is associated with kinetic peculiarities of crack propagation. Start of a crack in tension begins under attainment of critical coefficient of stress intensity K_{ic}, which gives rise to further avalanche crack propagation till full body fragmentation. The crack initiated under compression at K_{ic} is capable to grow uniformly under increase of continuous load only on a curved trajectory and tends to take its orientation toward the compression axis (Fig. 4.10).

Eventually, the transition of equilibrium crack propagation into avalanche crack stage, leading to full fragmentation is made possible after interaction of equilibrium growth of cracks at loads many times higher than the initial load for crack start. Experiments and theoretical analysis [6, 14] show that the total body fragmentation arises under the combined development of the interacting cracks after some equilibrium growth, as the growth of single crack cannot cause fracture even at infinite load. The steady creep rate of carbides does depend on kind of loading (compression, bending, tension) at temperatures $T > (0.65–0.70)\ T_m$ [22].

Table 4.4 Strength variation of refractory compounds at various loading modes (T = 280 K)

Kind of sample, material	Stress condition[a] base (mm)	Test number	$\sigma_m/(\sigma_{min}-\sigma_{max})$ (MPa)	$\Delta\sigma$ (MPa)	W (%)	m
Cylinder, ZrC,	I, 80	703	195/(105 – 135)	43	22.0	4.0
d = 3 – 4 mm, P = 7 %	II, 15	603	217/45 – 345	47	21.7	4.2
d_g = 9 – 20 μm	III, 50	60	80/57 – 129	18	22.2	4.3
	IV, 4	40	97/36 – 158	28	29.3	2.6
	V, 6	36	920/230 – 1240	24	26.0	2.5
Cylinder, d = 3 mm	I, 80	256	215/45 – 345	56	26.0	3.8
NbC, P = 8%, d_g=15 μm	II, 15	340	210/145 – 375	50	21.1	5.0
ZrC+5wt%C, P = 20 %, d_g = 15 μm	I, 80	120	85/28 – 220	43	51	2.3
ZrC, P = 75 %	VI	35	6.5/2.8 – 8.1	1.3	20.2	
ZrC+50%NbC, P = 65	VI	48	9.1/2.3 – 13	2.2	24	

[a]*I* four-point bending, *II* torsion, *III* tension, *IV* diametrical compression, *V* compression, *VI* hydrostatic tension (hollow cylinder d = 30 – 50 mm, ∂ = 4 mm, *H* = 50 mm)

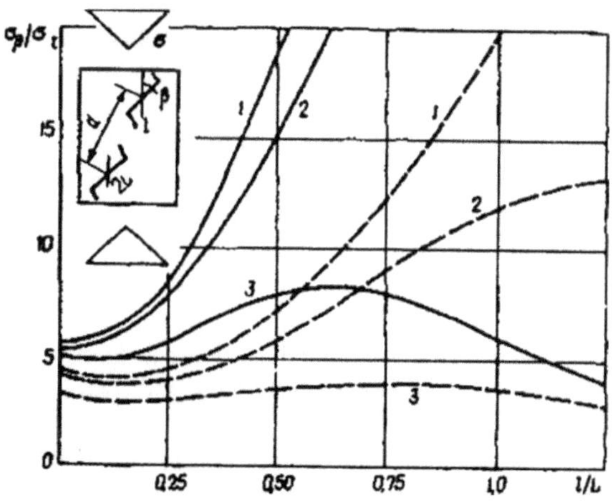

Fig. 4.10 Fracture diagrams for a system of parallel cracks: (*1*) $\lambda = d/2L = \infty$; (*2*) $\lambda = 3$; (*3*) $\lambda = 2$ (*solid lines* $\beta = 30°$, *broken lines* $\beta = 45°$)

The variation of loading conditions radically altering the strength level has no influence on the variation coefficient = S/σ and Weibull coefficient m (Table 4.4), where S is a root mean square strength deviation and am is a mean arithmetic strength value. The W and m are in the range 20 % and 3–5, respectively, for monophase dense and porous ceramic. The W value increases by two times for heterophase carbides with carbon inclusions owing to damage of carbide matrix. It is significant that the strength variation parameters of various ceramics in the brittle state are not affected markedly by electron band structure and atomic bonding but are governed primarily by the variation of surface and volume flaws.

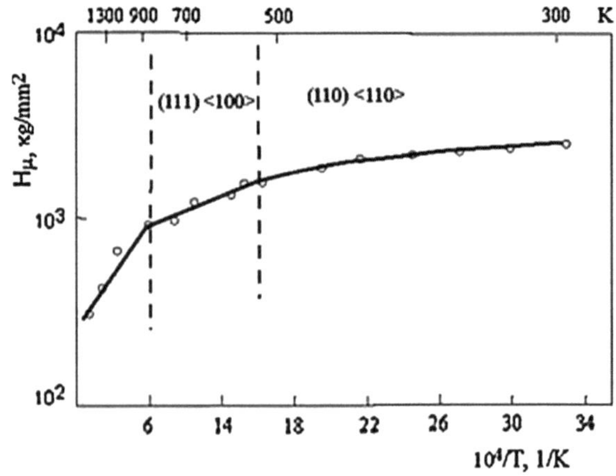

Fig. 4.11 Microhardness of monocrystal ZrC under various loads in relation to temperature with two distinctive changes

Transition from axial compression to multiaxial compressive loading by intrusion of indenter initiates the ductile deformation in carbides of transition metals even at 80 K. Temperature dependence of microhardness reveals deformation peculiarities (Fig. 4.11), undetectable under other kinds of loading.

The variation of loading conditions radically altering the strength level has no influence on the variation coefficient $W = \Delta\sigma/\sigma$, and Weibull coefficient *m* (Table 4.4), where $\Delta\sigma$ is a root mean square strength deviation and σ_m is a mean arithmetic strength value. Between distribution parameters *m* and W there is a correlation. Comparative tests of large and small sets of samples have shown that reliable enough estimation σ_m can be obtained from a test of 5–7 samples, and distribution parameters σ_n, W and *m* on 25–30 samples. The W and *m* are in the range of 20 % and 3–5, respectively, for monophase dense and porous ceramics. The W value increases by double for heterophase carbides with carbon inclusions, owing to damage of carbide matrix.

Transition from axial compression to multiaxial compressive loading by intrusion of indenter initiates the ductile deformation in carbides of transition metals even at 80 K. Temperature dependence of microhardness reveals deformation peculiarities, undetectable under other kinds of loading.

The first bend on the curve of microhardness takes into account the transition of gliding system {110} <110> to the system {111} <100> (Fig. 4.11). This is typical for transition metal carbides [20, 21]. The second bend associated with further development of ductile deformation and disappearance of cracks near indentations. The variation of load on indenter from 200 to 1,000 g changes the temperature of crack (c) disappearance from 800 to 1,400 K. The dislocation configuration around indentation (d) and elastic distortion zone (A), measured by Berg–Barret method, changing with orientation (Fig. 4.12), give valuable information about evolution of the deformation mechanism. The ratio of elastic zone (A) to the length of dislocation scatters (L) decreases by two times with temperature rise from 300 to 1,080 K through dislocation motion and relaxation of elastic stresses. Observation for changes

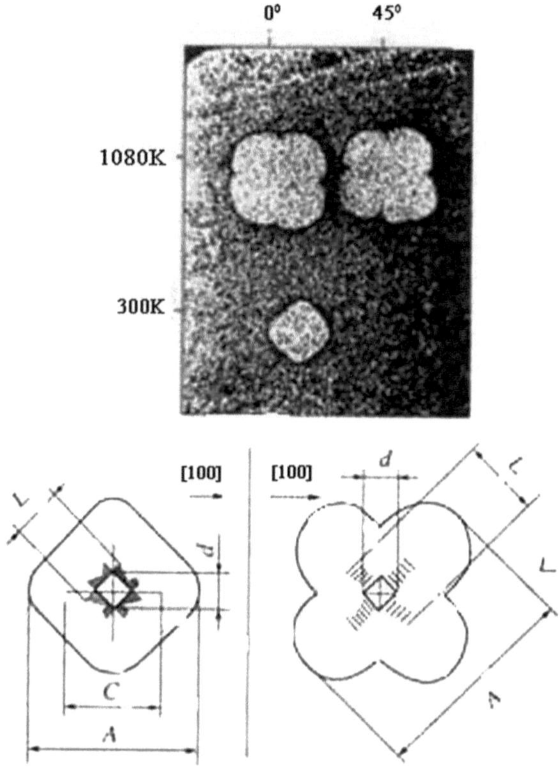

Fig. 4.12 Change of dislocation scatter elastic zone A, crack length C near indenter print d in monocrystal under 0 and 45° at 300 and 1,080 K

300K					1080K			
P, g	A/d	L/d	A/L	A/C	P,g	A/d	L/d	A/L
200	6.7	2.1	3.2	1.5	200	5.0	2.9	1.7
100	7.6	2.3	3.3	1.8				
50	8.5	2.0	4.3	2.3				

of elastic distortion and dislocation motion after annealing of carbide samples with indentations at various temperatures permits to determine the starting stress σ_s, for dislocation motion and yield stress σ_y [20]. The σ_s for ZrC and other refractory carbides is high in the low temperature range of 0.15–0.3 T_m as in covalent crystals (Ge, Si) with high Peierls stress (intrinsic lattice resistance to dislocation motion). σ_y is three orders of magnitude higher than σ_s. This suggests the controlling rate of dislocation generation and dislocation pinning. The value of as for metals is three orders lower than in ZrC and Ge. At temperatures above 0.4 T_m, the deformation mechanism changes and diffusion rate of metal and carbon atoms increases markedly. An active nonconservative motion of dislocations causes the relaxation of local stresses and subsequent decrease of yield stress. Elongated dipoles and dislocation loops disappear and generation of dislocation networks begins. The further temperature increase up to 0.6 T_m brings into existence the cell appearance. The formation of dislocation in carbides and metals during deformation are very similar. However, the temperature levels for initiation of dislocation motion and formation of cell structure are higher for carbides.

The fuel and structural NRE materials prepared predominantly by the methods of powder metallurgy have many structural defects, since the level of their strength is quite uncertain; varying within 15–25 % nears the mean value. The strength level depends on the loading method; but the variation coefficient W is in fact independent of the loading method. Changes in the strength of single-phase and heterophase carbide materials in passing from tension to bending calculated by the Weibull method are in good agreement with experimental data. This circumstance is taken into account in determining the reliability of mechanically loaded NRE constructions.

Dispersion of strength in an engineering practice most often is appreciated by Weibull distribution function:

$$P(\sigma) = 1 - \exp\left\{\int [(\sigma - \sigma_n)/\sigma_0]^m \, dV\right\}$$

where $\sigma \geq \sigma_n$ and σ_n is a stress below which the probability of fracture is equal to zero, irrespective of the dimensions of a body; σ_0 is the normalizing parameter; m is the parameter describing uniformity of a material, i.e., degree of distribution uniformity of defects on the body's volume. The distribution of stress estimated on the basis of a bend-and-torsion test of hundreds of samples of ZrC and NbC submits to the normal law and is described conveniently by the Weibull function.

4.3.2 Influence of Structural Parameters on Strength and Fracture

The strength of ceramics is dispersed in brittle condition in considerably greater degree than the strength of metals. This feature is defined, first of all, by variation of both superficial and volumetric deficiency and lack of stress relaxation on stress concentrator's influence on structural strength parameters.

The strength of ceramics in brittle condition is defined predominantly by the surface and volume defects [3, 14] and an alloying with formation of solid solution has practically no influence on the strength. Severe surface relief on sintered ceramic samples or microcracks appearing after electromachining or diamond cutting is responsible for the low strength level [14]. Elimination of these surface defects by mechanical grinding and polishing increases the strength by 50–70 % leaving the strength variance practically constant at the expense of volume flaws. The volume defects in the form of large rounded flaws with sizes of 100–400 μ are responsible for 40 % of all the cases of fracture in monophase ceramics. In some cases, the fracture occurs at lower stress at the cost of zone peculiarities appearing during inhomogeneous forming and subsequent inhomogeneous shrinkage during sintering.

Severe surface relief on sintered ceramic samples or microcracks appearing after electromachining or diamond cutting is responsible for the low strength level [14].

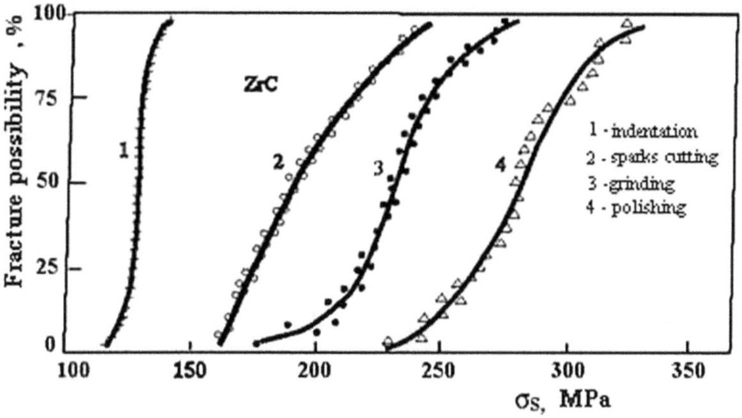

Fig. 4.13 Influence of surface sample defects of $ZrC_{0.97}$ after an electroerosive cutting- (*2*), grinding (*3*), polishing (*4*), and indentation (*1*) on the average level of strength and extent of data dispersion [14]

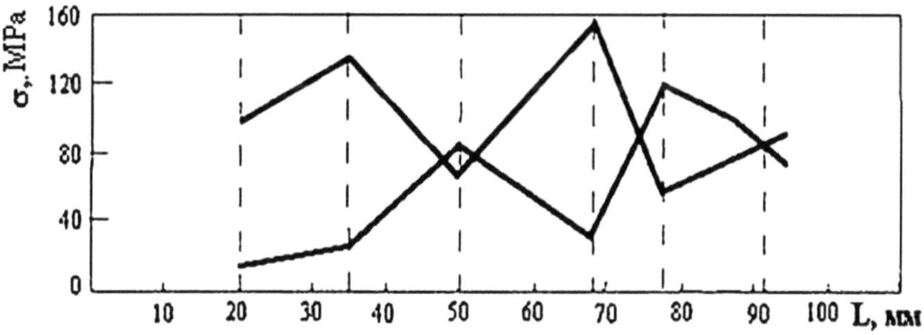

Fig. 4.14 Origination of residual stresses on both lobes of twisted fuel elements ZrC+UC owing to structural and chemical inhomogeneity in groups of grains

Elimination of these surface defects by mechanical grinding and polishing increases the strength by 50–70 % leaving the strength variance practically constant at the expense of volume flaws (Fig. 4.13).

Introduction of more dangerous stress concentrators on a surface, than in volume of the sample, by an indentation, reduces the average strength level but decreases dispersion sharply.

The dangerous defects in the form of cracks should be estimated by an X-ray method based on a measurement of a broadening of X-ray lines at the application of load to the sample, placed in a special prefix to a diffractometer [23] (Fig. 4.14).

The volume defects in the form of large rounded flaws with sizes of 100–400 μm are responsible for 40 % of all the cases of fracture in monophase ceramics. In some cases, the fracture occurs at lower stress at the cost of zone peculiarities appearing

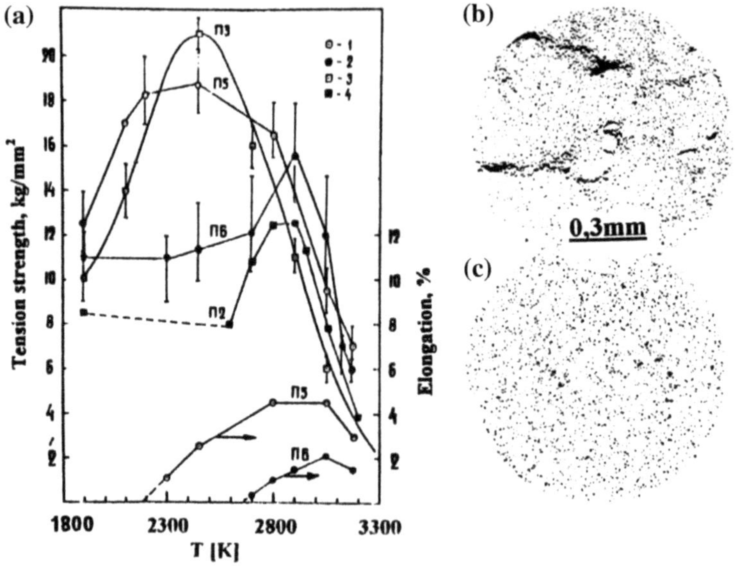

Fig. 4.15 Strength and deformation of ZrC with zone peculiarities (П2, П6) and without them (П0, П5) under tension in vacuum with deformation rate $3 - 10^{-3}s^{-1}$ (**a**). Microstructure of ZiC with zone peculiarities: (П2) (**b**) and without them (П3) (**c**). Batches of ZrC are manufactured by cold rolling (П2, П3) and pressing (П5, П6) with following sintering at 2,800 K in argon atmosphere 1 h [12]

during inhomogeneous forming and subsequent inhomogeneous shrinkage during sintering (Fig. 4.15).

These zones in the form of friable porous layers lower the strength by 30–50 %. Maximum of strength and T_{b-d} is increased by 300–400 K in comparison with ceramics having the same porosity level but without these zones [12, 14]. In most cases, sources of failure of ceramic materials are small cracks faintly visible at the magnification of 1,000 with a radius many orders of magnitude lower than for large pores; so, the former are able to cause more dangerous stress concentration.

The availability of pores decreasing the body's cross-section clearly reduces the strength. The pore size increased from 3 to 90 μ and emergence of elliptical pore form at a constant porosity level led to strength reduction of ZrC under bending from 300 to 170 MPa with constant porosity level of 5–7 %. The exact influence of porosity on the strength is difficult to determine as in most cases, simultaneously with porosity, other structural parameters are varied: grain size, surface and volume defects, and impurity segregation [3].

Structural parameters depend to a great extent on the chosen technological parameters. The presence of pores in a material naturally reduces the cross-section of a body and its strength, which can be described, for example, by an empirical relation like

$$\sigma = \sigma_0 \exp(-BP),$$

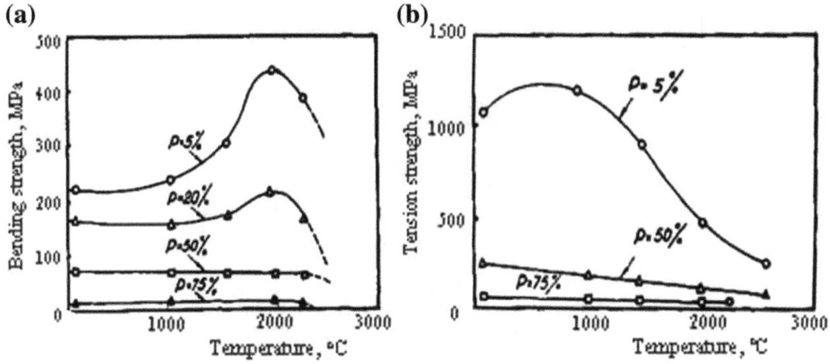

Fig. 4.16 Temperature strength dependence of zirconium carbides at bending (**a**) and compression (**b**) with various porosity at deformation rate $3 \cdot 10^{-3}s^{-1}$

where σ_0 is the strength of a body without pores and B is a coefficient depending on the pore size and configuration [24]. The temperature of brittle–ductile transition of ceramics versus porosity increases. For example, T_{b-d} of ZrC, NbC at raise of porosity from 7 to 60 % increase T_{b-d} by 400 K at a bend test and by 600–700 K at compression (Fig. 4.16). The maximum strength, owing to suppression of a plastic deformation is decreased and shifted toward to higher temperatures. Similar changes occur at carbide–graphite compositions.

The relationship between strength and grain size is rather complicated due to simultaneous variation with grain size of flaws, segregation additives on the grain boundaries and ratio of grain and boundary volumes in ceramics [2, 3, 14] (Fig. 4.16).

The primary recrystallization (annealing of material after preliminary deformation) makes it possible to vary the grain size of ZrC in the wide range from 5 up to 2,500 μ without altering the boundary conditions. In this case, the variation of grain size retains carbide strength at 280 K constant while a decrease of strength of ZrC samples is observed after accumulative recrystallization (Fig. 4.17). The latter is more common for ceramics, as mentioned in references. The yield stress $\sigma_{0.2}$ of ceramics follows Hall–Petch law in the temperature range $0.6T_m < T > T_{b-d}$ [25].

$$\sigma_{0.2} = \sigma_0 + k_y dg^{-1/2}$$

where σ_0 is the Peierls stress and k_y is the coefficient of deformation resistance through grain boundaries. The relationship is altered radically at temperatures above $0.6\,T_m$. The yield stress $\sigma_{0.2}$ decreases owing to the grain boundary sliding and rotation of crystals.

Microstructural defects such as pores, grain boundaries, and microcracks located between grains or phase components influence the resistance to cracks for materials prepared by powder metallurgy methods. The value of K_{1c} for carbide single-phase materials does not exceed $3\,MPa\,m^{1/2}$. As a rule, K_{1c} no monotonically changes with temperature. First, K_{1c} decreases or remains unchanged with increasing the

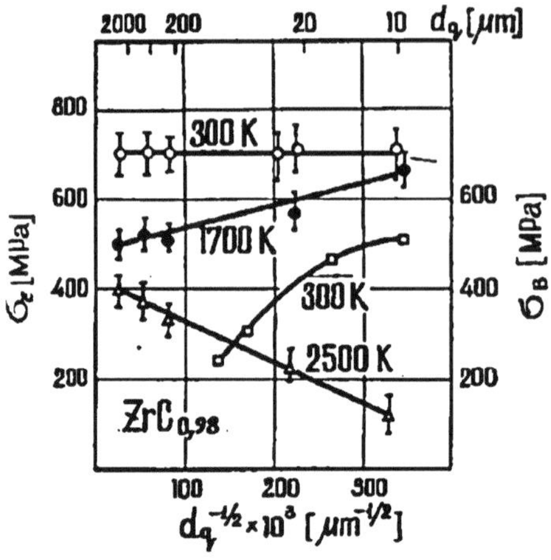

Fig. 4.17 Compression strength σ_c for ZrC in vacuum after primary recrystallization and bending strength σ_b for ZrC obtained by accumulative recrystallization against grain size

Table 4.5 Results of rejection of fuel elements from ZrC + NbC + UC

Loading mode	Level of loading σ_{rj} (MPa)	σ_{rj}/σ_m	Amount of fuel elements	Amount of rejection (%)	Strength gain (%)
Bending with rotation	125	0.6	300	5	5
	140	0.7	1,800	8	20
	175	0.85	450	35	23
Rotation	10	0.6	23,000	3	1.5

temperature and begins to increase only when plasticity develops. The effective surface energy depends on the environment in a complicated way.

For exclusion of the most defective work pieces and the decrease of strength dispersion, it is expedient to carry out rejection by a preliminary mechanical loading [6]. For achievement of a positive effect of the rejection, it is necessary for each product to install the mode and level of a loading on the basis of the statistical data about strength distribution taking into account a possible damageability and economic feasibility (Table 4.5).

4.3.3 Temperature Dependences of the Strength and Creep

The working temperature of materials in different parts of HGAs in NREs in transient and stationary regimes changes in a broad range from 300 to 3,000 K. since it is necessary to know the variations in mechanical and physical properties. For structural and fuel materials of the interstitial phase type, the problem of preventing a brittle collapse is of primary importance. Such collapse is caused by the high brittleness of materials of this class in a broad range of temperatures, loading methods and

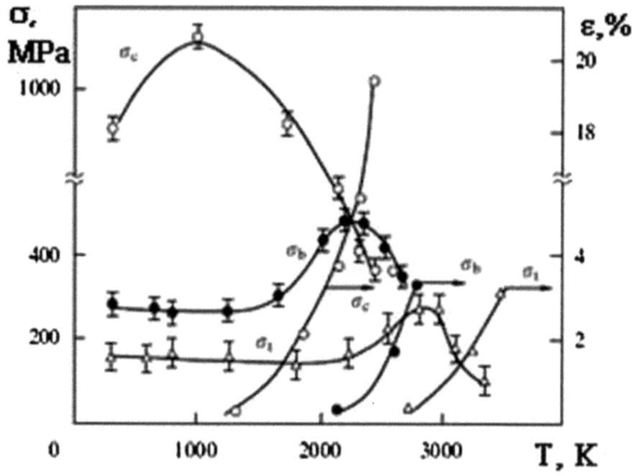

Fig. 4.18 Dependence of ZrC strength and relative deformation ϵ at various temperature according to kind of loading; σ_t, bending σ_b. compression c with deformation rate V = $5 \cdot 10^{-3}\,s^{-1}$ [14]

rates, along with a relatively low tensile strength (compared, for example, with metal materials). Refractory carbide compounds with a high level of resistance to the motion of dislocations are characterized by no monotonic dependence of the strength [14]. First, the strength is virtually unchanged with increasing the temperature and begins to increase only when microplasticity develops, achieving a maximum (Fig. 4.18) at the brittle–ductile transition temperature T_{b-d}, due to the thermally activated relaxation of local peak stresses near concentrators–structural defects contained in the material. As the temperature was further increased ($T > T_{b-d}$), microplasticity developed in ceramics and their strength monotonically decreased, as in metals. The presence of nonmetal (O_2, N_2, C, Si) or metal impurities (Fe, C, Ni) due to the formation of phases and low-melting eutectics can change the transcrystalline type of collapse to the intercrystalline one and also change the temperature dependence of the strength.

The brittle–ductile transition temperature in ceramics, unlike T_{b-d} in metals, strongly changes (by a few hundred degrees) upon changing the type of the stressed state and the deformation rate. As the deformation rate of ZrC rise under tension from 10^{-3} to $10^{-1}\,s^{-1}$, the temperature T_{b-d} increases from 0.6 to 0.8 T_m. In the case of shock loadings with rates exceeding the relaxation rates of micro- and macrostresses, the strength remains almost at the level of its value at room temperature. A similar behavior is also observed for oxide materials and weakly plastic Si and Ge metals with strongly manifested covalent bonds (Fig. 4.19).

Transition temperature T_{b-d} for complex solid ZrC + UC, ZrC + NbC + UC solutions is higher than for the single-phase ZrC.

Long-term strength with increase of loading time drops. Time of fracture τ and stationary creep rate ε' are connected with applied stress σ. Laws of creep kinetics inherent to metals are observed and for ceramic materials [22, 26]. In a general view dependence of creep rate ε' on applied stress σ can be written down:

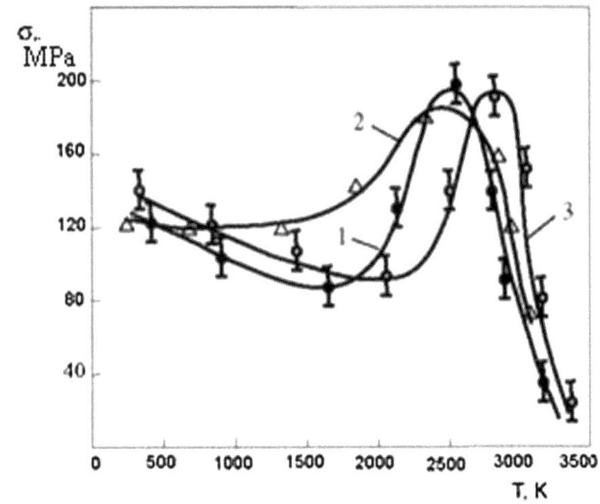

Fig. 4.19 Temperature dependence of tension strength σ. 1-ZrC0.9+5%UC; 2-NbC0.9+5%UC; 6-ZrC+35%NbC+15%UC

$$\tau = B \cdot \sigma^m \cdot \exp(U/RT)$$
$$\varepsilon' = A \cdot \sigma^n \cdot \exp(-Q/RT)$$

where A, B, n, m—constants, Q and U—activation energy of creeps and long fracture processes. For carbide materials tested in a temperature band 2,500–3,000 K, activation energy Q and U and exponent's n and m of stresses practically coincide.

Creep rate depends from σ mostly under the linear law in an interval of small stresses 10–20 MPa at high temperatures $0.6\,T_m$. According to this, we usually distinguish threshold (i.e. developed at stresses surpassing some stress $\sigma \geq \sigma_t$) and not threshold creep; the last can be observed at any as much as small σ. No threshold creep is carried out by the directed diffusion of atoms and proceeds on vacancy–diffusion mechanism of Nabarro–Herring. Creep rate inversely proportional to a square size of grain and in a strong degree depends on stochiometric of the compound (Fig. 4.20a).

At stresses $\sigma > \sigma_t$ alongside with diffusive processes, conservative movement of dislocations is initiated and dependence ε′ from becomes power mode $\varepsilon' \sim \sigma^{(3-4)}$ (Fig. 4.20a) and the intergranular sliding at the final stage of creep generate microporosity on the grain boundaries that causes appreciable decrease of density. Generally, creep deformation is caused by passing not one but several processes (diffusive fluidity. sliding on grain borders, dislocation moving), the contribution of each process depends on conditions of carrying out of experiences (temperature, stress, duration) and structural characteristics of the object.

The instantaneous and unsteady stages of creep are absent in most cases at stresses greater σ_t, but smaller yield point $\sigma_{0.2}$. Accumulation of a plastic strain basically occurs at stages of the stationary σ_t and accelerated creep, and duration of a final site leaves approximately 1/3 from time before fracture. The general accumulated plastic strain during creep

$$\varepsilon = \varepsilon'\tau + \varepsilon_{III}(\tau)$$

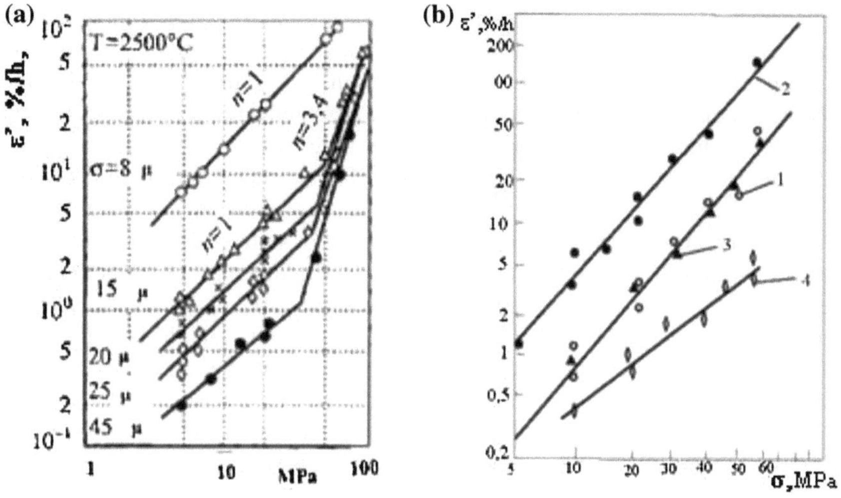

Fig. 4.20 Dependence of creep rate of ZrC with various grain value versus stress at T = 2,900 K (**a**) and (**b**) for fuel materials 1-ZrC–UC, 2- ZrC-UC-C, 3- ZrC-UC-7mass.%NbC, 4-ZrC–UC-48mass %NbC T = 2,800 K

where ε'—creep rate on the stationary stage; ε_{III}—deformation on the accelerated stage; τ—time.

The structure of ceramics changes with increase of a creep strain, agglomeration of microspores in the form of chains on borders occur in grains, perpendicular to action of loading. Structural changes result in to change of density and physical properties. Increase of porosity, for example, in beryllium from 1.4 up to 12 % raises ε' in 2–4 times. Generally, porosity dependence of creep can be expressed by empirical expression.

$$\varepsilon' \sim (1 - p2/3)^{-1} \quad \text{or} \quad \varepsilon' \sim (1 - p)^{-3,}$$

where p is porosity of a sample.

The fatigue behaviors of zirconium carbide [14] under cyclic loading have demonstrated that fatigue cracks in ceramics can grow at lower stress than static strength. The cyclic tests were carried out under compressive loading at room temperature in ambient air. Subcritical crack velocity is described by:

$$\mathrm{dl/dN} = \mathrm{a}(\mathrm{K_i}/\mathrm{K_{ic}})^{\mathrm{n}}$$

where K_i is a current stress intensity. K_{ic} is the fracture toughness, a and n constants depending on type of loading and the environment, K_{ic} is determined by the developed method under uniaxial compression for plane specimens with a central cut at a certain angle to an axis. The crack starts at K_{ic} and after sporadically increasing length stress intensity drops to a lower K_{1a} value. The subsequent crack growth is possible only at increasing external load, these permits to multiply determination of the fracture toughness on one specimen. The results of cyclic tests show that crack propagates even at $K_i < K_{ia}$ with velocities 1–3·10^{-3} mm/s. The value dl/N increases to 1 mm/s at $K_i > K_{ic}$, and *n* value in the range 75–89. The observed subcritical crack growth under compression at room temperature is connected with accumulated localized microcracking and crushing of fracture-surface roughness at the unloading stage. Internal local stresses play an important role in the fatigue behavior of ceramics. The origin of these local stresses depends on heterogeneity of chemical composition, structure, and anisotropy of sintered ceramic blanks [14]. Cyclic loading of cylindrical ZrC samples at room temperature does not decrease the strength σ_b under constant bending stresses equal to 0.8–0.9 of the mean, σ_b for a number of cycles up to 10^6. The cyclic bending of ZrC samples at temperature above 2,300 K increases the strength because of viscous intergranular flow and stress relaxation of local stresses.

The resistance of ceramics to the propagation of cracks can be strongly influenced by the microstructure but optimization of fracture toughness and strength usually involves different choices of microstructure: individual toughening mechanisms including phase transformations. Microcracking, twining, ductile reinforcement, fiber/whisker reinforcement and grain bridging as regarded thoroughly in [3]. The optimum structure of ZrC with small grain, pore size below 1 μ and medium porosity of 8–10 % has the maximum $K_{1c} = 3.3$ MPa·$m^{1/2}$ and bending strength near 650 MPa [14]. The K_{ic} of most refractory carbides, nitrides, and borides does not exceed 5 MPa·$m^{1/2}$ [2]. The polycrystals based on two modifications of boron nitride with large internal local stresses have extremely high K_{ic} up to 18 MPa·$m^{1/2}$ [2].

4.3.4 The Thermal Stress Resistance

The thermal stress resistance *(TSR)* is the most important characteristic of the efficiency of HRA construction elements [6, 14, 27, 28].

A thermal elastic stress can be easily determined at known temperature patterns from combined equations of stress and strains with consideration of thermal deformation. Dependences between shear stresses and strains, and also the static and geometrical equations are accepted without modifications on theory of elasticity [24, 29].

The thermal stresses σ can be defined by following relationship:

$$\sigma = E \cdot \alpha \cdot \Delta T_m \cdot K_f/(1 - \mu),$$

where $\Delta T_m = (T_m - T)$ is a calculated mean integrated temperature difference $\Delta T_m = (T_m - t)$ at the moment of fracture (where T_m is a mean integrated temperature of a body cross-section and T is surface or center point temperature, assuming that properties of a material do not vary and material is elastic), α is a coefficient of thermal expansion, E is a modulus elasticity, μ—is Poisson's ratio, K_f is form factor. Derived from the relationship at value $K_f = 1$ complex $\sigma(1 - \mu)/E \cdot \alpha = \Delta T_m$ is accepted to

Table 4.6 Thermal strength R of products made of different carbide materials in the elastic–brittle state[a] (T = 600 − 1,200 K) [29]

Product	Material	α 10^6 (K^{-1})	E 10^{-3} (Mpa)	σ_1 (MPa)	λ (W m^{-1} K)	R (K)	W (%)
Casings. Dout = 52 mm.	Carbon-fiber plastic ZrC, porosity $\rho = 75\,\%$	5.1	17	8	0.5	95.0	24.0
Wall thickness $\delta < 5 = 2.2$ mm	Porous ZrC, $\rho = 75\,\%$	6.3	16	11	3.0	68.0	21.0
	Porous ZrC + NbC, $\rho = 65\,\%$	5.5	30	22	3.0	72.0	15.0
	ZrC, $\rho = 5\,\%$	5.1	350	250	20.0	50.0	22.0
Cases. −Dout = 48 mm $\delta = 0.75$ mm	NbC, $\rho = 12\,\%$	4.8	310	260	35.0	43.0	14.0
	ZrC + NbC, $\rho = 10\,\%$	4.9	350	330	25.0	56.0	19.0
	ZrC 'flaking'	5.0[b]	3.0[b]	6.0	1.0[b]	>400.0	
Fuel elements $D_n = 2.2$ mm	ZrC+5.2%UC, P = 5 %	5.2	350	320	12.5	51	20
	ZrC+NbC+4.6%UC, P = 5%	4.9	370	395	10.0	72	19

[a]Data for other properties correspond to T = 600 K [b]For anisotrophic 'flaking' E and a are measured in the direction parallel to the deposition plane, while λ is measured in the direction perpendicular to this plane

Fig. 4.21 TSR for ZrC disk (30 mm in diameter with thickness 2–3 mm) as function of temperature and heating rate. (*1*) 10 K/s, (*2*) 50 K/s, (*3*) 100 K/s, (*4*) 200 K/s

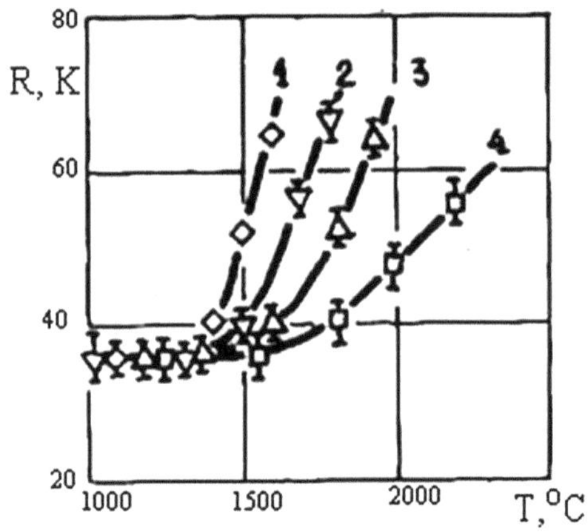

name the first criteria of thermal stress resistance $R = \sigma_t(1 - \mu)/\alpha_t E$ measured in degrees. The complex $R' = \lambda R$ is a second criteria, which reflect ability of materials to maintain a thermal flow at stationary or nonstationary thermal influences.

The higher the heat conductivity λ of a material is the greater the thermal flux that can be withstood by the body without its collapse. It should be noted that the choice of materials for heat insulators in constructions based on the value of R' is meaningless. In this case, it is necessary to use the value of R for estimates because, given two materials with identical heat-insulating properties, the better material is the one capable of withstanding a greater temperature drop.

The TSR of most refractory ceramics lie in the range of $R = 30-100$ K (Table 4.6). The criterion R' can change in more wide limits 100–1500 W/m. The thermal stress σ_{th} (causing appearance of first cracks) is slightly exceeds the tensile strength and σ_{th}/σ_b, lies in the range of 0.45–0.57 of bending strength value.

The knowledge of TSR versus temperature is necessary for the appraisals stability of ceramic articles under working conditions. The temperature dependence of TSR has three characteristic intervals [6]. The TSR of ceramics is invariable in the elastic–brittle temperature range. A relaxation of thermal stresses increases the TSR with temperature rise in the second interval. The tests performed on ZrC disks by induction heating method [6] show this possibility (Fig. 4.21).

The R value of ZrC remains constant with temperature increase up to 1,700 K regardless of heating rate. The TSR criteria R and R' of a porous ZrC considerably change when the materials porosity varies (Fig. 4.22). Pores reducing the cross-section area of a body cause a drop in strength σ and elasticity modulus E. The σ and E dependencies on porosity P may be described by empirical relations of this sort:

$$\sigma = \sigma_0 \exp(-BP), E = E_o \cdot \exp(-BP).$$

The invariability of TSR in the range from 500 to 1,700 K is conditioned exclusively due to the constancy of strength in this range, since the E decay and α rise

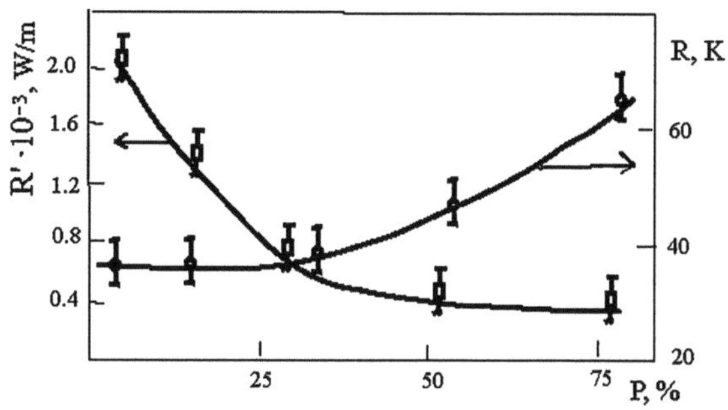

Fig. 4.22 The dependence of TSR criterion R and criterion R′ for ZiC as a function of porosity

with temperature compensate for each other and the product αE is kept practically constant. The lower is the velocity of thermal loading above 1,700 K the greater is the reduction of the local stresses and the greater is the up growth of TSR. The thermal fracture in the third temperature range cannot be observed when the thermal stresses relax more rapidly than their formation. The ZrC samples do not fracture even at 1,900 K at low heating velocities <10 K/s and much less at the onset of the macroductility at 2,000–2,200 K.

Distribution of R values for ZrC, NbC measured on a representative number of samples (70–90) is characterized by exponential Weibull relation. The coefficient of variation $W = S/R_m$ (S is deviation of TSR, R_m is average value) agrees closely with appropriate significance of the W value obtained under mechanical test of strength. The cyclic thermal loading does not decrease the carrying capacity of samples in elastic–brittle temperature range. TSR of the ceramic bodies can be seriously changed under combined influence of the thermal and mechanical loads [28].

TSR of the ceramic materials can be essentially raised by introduction of hard inclusions in a matrix. The introduction of second phases as carbon inclusions with the sizes varying from 0.5 up to 50 μ into ZrC and NbC matrixes inevitably reduces the strength but is capable to increase the fracture toughness by a factor 1.5–2 with an optimum choice of concentration and kind of carbon inclusions. With growth of the carbon content, the TSR either increased in case of use of carbon black or decreased for graphite [6]. TSR of compositions with carbon black can be raised 2.5–3 times (Fig. 4.23) due to an increase of relation σ /E [6].

The volume content of the inclusions, their geometry and dimensions have also a notable effect on the TSR as shown by ZrC-compositions with diamond particles [17].

Use of synthetic diamond powders with a narrow spectrum of grains get to obtain the compositions (after sintering) with guaranteed dimensions of the carbon. The inclusions are distributed in the matrix uniformly due to a good mixing of the ZrC and diamond powders. In addition, the diamond volume change during the structural transformation has an active effect on the process of the composition compaction when sintering. With introducing small carbon particles (0.6–6 μ) up to 5 mass % of the strength and the fracture toughness are not practically changed but the modulus of elasticity is decreased considerably. Introduction of 50 μ particles results in simultaneous drop of σ/E and K_{ic}. Inadequate change of the strength and the modulus of

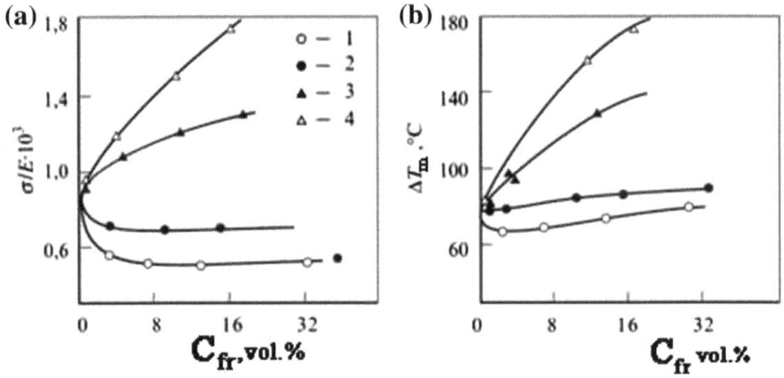

Fig. 4.23 Modification of values of σ/E (**a**) and (**b**) thermal stress resistance ΔT_m in composites based on zirconium and niobium carbides depending on the contents of the carbon component in the form of carbon black (*3, 4*) or graphite (*1, 2*)

elasticity when increasing the volume content of the diamond particles, results in the σ /E-value growth. Therefore, the TSR of the compositions with the diamond particles 0.6–6 μ in size is increased to a greater extent. Apparently, some increase of the fracture toughness does not influence the TSR considerably. The TSR of compositions with carbon inclusions is left invariable in temperature range 400–1,900 K and the advent of the first evidence of macroductility happens at more higher temperature up to 2,400 K.

4.4 Materials of the Heat Insulating Package

Researchers of RIPRA "Luch" have invested considerable efforts into the development of various alternative manufacturing techniques for processing heat insulating materials (HIP) of the nuclear fuel assemblies (NFA) of Nuclear Rocket Engines (NRE), shows two alternative packaging arrangements of heat insulation packages (HIP) for NFA.

Five-layer HIPs based on three ZrC + C casings Fig. 2.9b are designed for the thermal flow values of $2 \times 10^6\,W{\cdot}m^{-2}$ [30]. The HIP protects the NFA housing from the heat impact of the working medium. This design features a multilayer sectional package structure that minimizes the possibility of cracks penetrating into the housing and offers the possibility to vary HIP material composition over its length and across its thickness.

The pyrolytic graphite-based outer casings of the HIP, along with good thermal insulation, provide for a 'soft' contact with the housing, thereby facilitating the structure assembling process and minimizing abrasion of the metal case by thermal insulation. The inner casings are thin-walled carbide–graphite cylinders. For the low-temperature section these casings are made of zirconium carbide + graphite (ZCG) and for the hot section the niobium carbide + graphite (NCG) casings are used. These casings serve as a supporting frame for the HRA preventing the penetration of fragments of heat insulation elements into the hot sections' duct. The casings ensure the assembly of HSs and their mounting into a heat-insulating package and reduce the effects of erosion and aggressive impact of the working medium on the heat insulation. Casings made of low-density pyrographite and porous ZCG and NCG and typically placed between metal housings. Casings made of low-density pyro-graphite are located in the low-temperature section ($T = 1{,}500 - 2{,}000$ K). For higher temperatures, the first designs contained casings based on so-called "laminate" consisting of carbide layers in a graphite matrix. The next-generation designs had these replaced by casings made of porous Zr and Nb carbides.

The HIP internal casings are made of ZCG or NCG containing 5 wt% of the free carbon. Use of ZCG is limited by the eutectic melt temperature of compositions. The NCG layer serves as a rigid carcass preventing the possible penetration of HIP fragments into the hot section duct. The low-density pyrolytic graphite layers (deposited on the surface of one or two casings) are used in the zones characterized by the maximum power flux. The length of this zone is limited by the temperature of reaction with the environment (hydrogen).

Preforms of ZCG-based on laminate pyrolytic compositions (ZrC + Pq) were processed by depositing carbon and zirconium particles from the vapor-phase mixture of $ZrCl_4\,CH_4 + H_2 + Ar$ on the surface of a thin-walled graphite substrate at 1,700–1,800 K [31].

Pyrolytic graphite (PGV) is a polycrystalline form of graphite deposited from a vapor-phase carbon source at pressures of 25–150 mm Hg on a heated graphite substrate [32, 33]. The PGV casing is used in the relatively cold zone of HIP. PGV casing provides reliable heat-shielding of the duct housing and prevents metal housing exposure to hydrogen. PGV has a highly desirable feature for the given operating conditions: its thermal conductivity slightly decreases with temperature rise.

Pyrographite properties depend primarily on the substrate temperature, pressure and speed of the flowing vapor and on the composition of a vapor mix. By rising the deposition temperature from 1,800 to 2,500 K it is possible to obtain pyro-graphites with the density varying from 1.2 to 2.25 g/cm^3.

Such density variations lead to significant changes in the physical properties of the pyrographite; thus the density is considered as the basic parameter influencing the main properties, including thermal conductivity. Distinctive feature of pyrographites is their significant anisotropy of properties. In particular, perpendicularly to the deposition surface (*c* axis) thermal conductivity may be 50–100 times lower than that along deposition surface (*a* axis); depending on density; the thermal conductivity may vary from 1.5 to 7 W/m·K [33]. Due to the lower thermal conductivity along c-axis pyrographites may effectively be used in heat shields for various power plants with high operation temperatures.

The results of analysis of the pyrographites thermal conductivity values show considerable scatter both in absolute values, and in their change with temperature along c axis [31–33]. This further complicates thermal calculations and the prediction of performance of specific thermally stressed parts of HGA. In order to obtain the

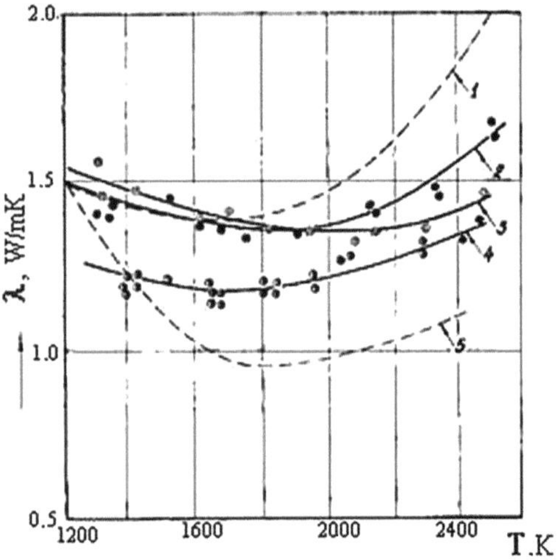

Fig. 4.24 Heat conductivity of high density pyrolitic graphite PGV measured on the ring samples (*1–4*) which have been cut out from regular products. *Dashed lines* (*5*) are the literary data [32]

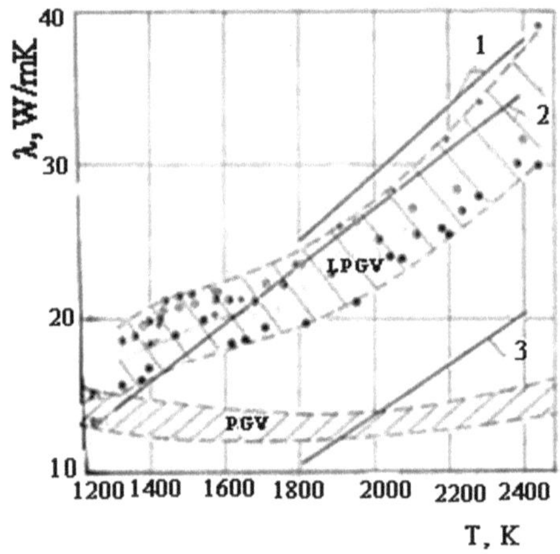

Fig. 4.25 Temperature dependence of heat conductivity of casings from low density LPGV with various density 1.35–1.60 g/sm^3 and PGV

reliable data on thermal conductivity of pyrographites, measurements were made on samples processed by standard techniques routinely used for production of standard NFA parts. PGV thermal conductivity does not change monotonously. As temperature rises within the certain range, the thermal conductivity decreases reaching a certain minimum and then rises again (see Fig. 4.24).

The maximum difference between absolute values of heat conductivity for the samples performed at temperatures ~1,300 K does not exceed 20 %, while for those made at ~2,400 K it does not exceed 13 %, which is quite acceptable for the pyrolytic graphite processed by the same technique. The measurement error λ for pyrolytic graphite in all cases does not exceed 8–10 %.

Scatter of thermal conductivity values for standard high-density pyrographite as well as the one of the literature values is attributable primarily to multiple defects, as

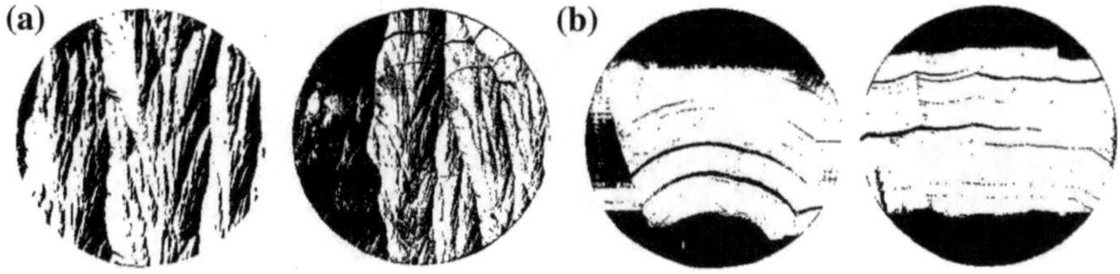

Fig. 4.26 Structure of pyrolitic graphites PGV (**a**) and low density PGV (**b**) (magnification x70)

well as to micro- and macrostructural differences in PGV parts (see Fig. 4.24) caused by differences in processing history (deposition temperature, heat treatment, gas pressure and flow), sample geometry and other factors. PGN thermal conductivity is typically higher than that of PGV deposited at 2,300–2,500 K (Fig. 4.25). Such change in thermal conductivity with density is due to the structural changes in the material, this assertion being confirmed by the X-ray structural analysis and microstructural characterization.

PGV structure is typically characterized by presence of a significant amount of microcracks (see Fig. 4.26). The reasons for microcracking being (1) stresses generated by temperature gradient across the layer during cooling, and (2) anisotropy of thermal expansion along c and a axes. The thicker the PGV layer, the greater are the stresses and consequently, the higher the propensity to microcracking.

The standard PGV products with wall thickness δ of 0.75 mm may feature microcracks of considerable length that typically pass through multiple growth cones. Upon reaching the cone the crack stops, and the new one is generated in the adjoining layers. Virtually defect-free structure is highly unusual in standard PGV products. As the microcrack plane runs nearly parallel to the deposition surface such microcracks act as large thermal resistances during heat transfer by thermal conductivity in the direction perpendicular to the deposition surface. The number of microcracks has a considerable effect on the PGV thermal conductivity value and on its behavior with temperature change.

Powder metallurgy technique of processing the heat insulating materials has a considerable effect on their structure and properties [10, 27]. Porosity increase in ZrC-based samples from 5 to 70 % leads to thermal conductivity drop by an order of magnitude depending on the structural features, particularly, on skeleton structure factor and closed or open porosity.

For example, porous fibrous ZrC (see Fig. 4.27d) processed by thermal diffusion impregnation of carbonized fiber by zirconium show lower thermal conductivity than carbides with continuous skeleton matrix, despite their equal density. Natural structural variation in a porous composition inevitably leads to the scatter in heat conductivity values between samples from the same batch. It is to be noted that the temperature coefficient a thermal conductivity in highly porous carbide materials is similar to the one for the dense carbides, with some influence of heat radiation in pores noticeable only at temperatures above 1,800 °C.

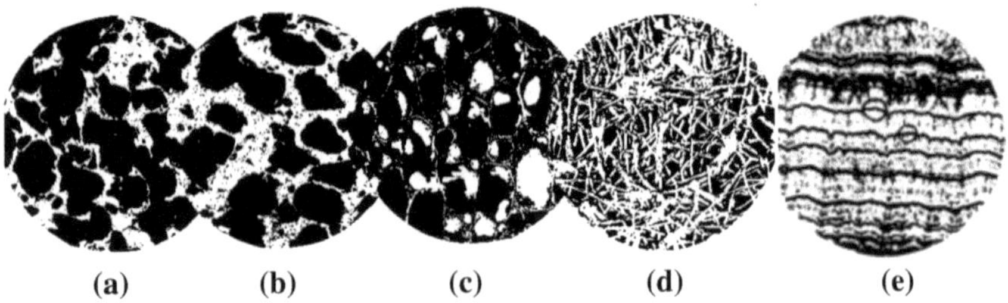

Fig. 4.27 A microscopic structure of porous ZrC carbides (**a, b**), foam ZrC carbide (**c**). fibrous ZrC carbide (**d**) all with porosity (P ≈ 75 %) layered carbide–graphite (ZrC + Pq. P = 25 %) (**e**), (**a, b, c** x200), (**d** x70)

ZrC-based fibers with nearly stoichiometric composition had diameters of 15–20 μm. Bulk density of carbide, depending on a carbonization regime, varied from 70 to 90 % with thermal conductivity ~2 W/m·K at porosity ~70 %. Compressive strength values of these samples are rather low only 1.2–1.5 times that of tensile strength [33]. At equal porosity levels, porous carbides are stronger than fibrous ZrC, though demonstrating somewhat lower thermal insulating capacity than the latter. Greater shrinkage of fibrous ZrC at temperatures ≥2,300 K even under moderate compressive loads makes porous carbide the material of choice for use in HIPs. Greater structural stability of porous carbide, attributable to its structure, is influenced by the choice of processing technique [34]. Zirconium carbide and potassium bromide powders in ratios 1:10 to 1:300 were mixed with binder (natural rubber). This premix was pressed and pre-sintered in vacuum for de-bonding at temperatures above carbide sintering point, with subsequent final sintering to achieve consolidated porous parts. Pre-sintering was effected with heating ramp up to 2,300 K during 11 h, with subsequent isothermal dwell time of 0.5 h, whereas final sintering was carried out at heating ramp up to 2,800 K within 5 h with subsequent dwell time of 0.5 h at maximum temperature. This sintering technique yielded products with porosity of 50–85 %, depending on the introduced filler amount. Pore stability at high temperatures was achieved through the presence of rather large cellular pores corresponding to the pore-forming agent (filler) particle size.

Formation of special structure of a heterophase composition from alternating carbide–graphite layers (ZrC+Pq) with porosity of 30 % (Fig. 4.27e) deposited from a gas phase get to gain stronger material with thermal insulating properties (Fig. 4.28), which are not conceding to properties of high porous carbides.

It should be note that the temperature coefficient of thermal conductivity of high porosity carbide materials is analogous to that for the dense carbides (Fig. 4.29), some influence of heat radiation in pores are detected only at temperatures above 1,800 °C.

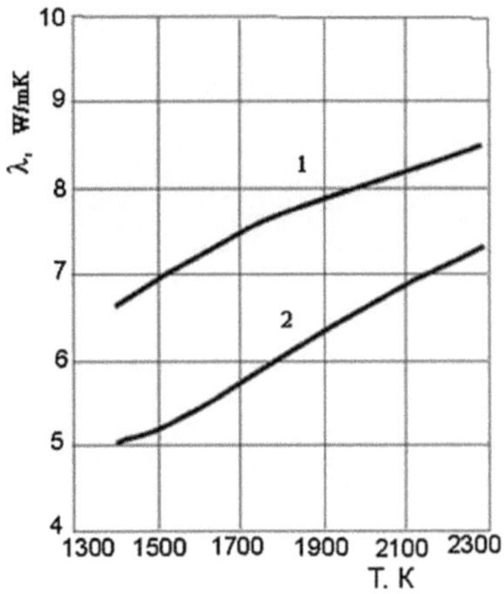

Fig. 4.28 Heat conductivity of compositions ZrC+Pq from alternating carbide–graphite layers: ZrC 35–40 weight.%, PGV 60–65 weight%, (*1*) and ZrC 55–60 weight%. PGV 40–45 weight% with a thickness of 0.25–0.5 mm every layer (*2*)

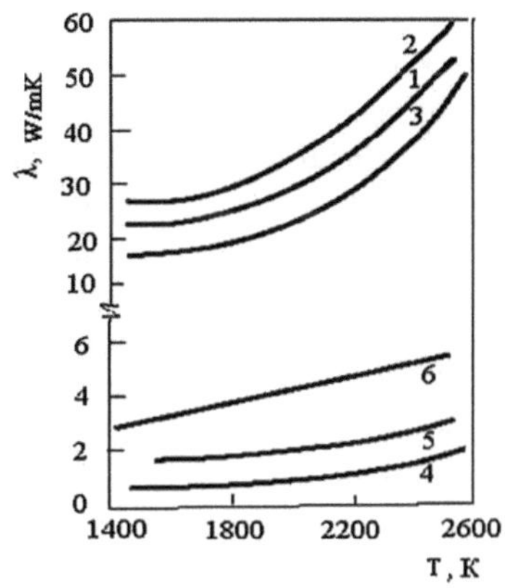

Fig. 4.29 Heat conductivity of ZrC, П = 6 % (*1*), ZrC +5%C (2), ZrC +5%UC (*3*) pyrolitic graphite (*4*), fibrous ZrC carbide p = 70 % (*5*), foam carbide ZrC П = 65 % (*6*)

4.5 Hydride Compositions for a Neutron Moderator

Transition metal hydrides may be used as neutron moderators for small-scale highly stressed nuclear power installations (NPI) at higher temperatures [7, 35]. Zirconium hydride moderator blocks of NRE should provide structural integrity during thrust regime and remain chemically inert under bimodal power regime at power fluxes of 1–1.5 kW/m^3 at average temperatures of 570 K and temperature gradients of ~110 K across the block.

Zirconium hydride parts are typically processed by reach-through hydrogen impregnation of the heated metal preforms, or by molding metal hydride powders with subsequent pressure treatment [36]. The terms "reach-through impregnation" and "reach-through hydrogenation" mean diffusion-controlled impregnation of metal preforms up to the certain hydrogen volume fraction. Important advantage of a compacted preform hydrogenation method is the possibility of processing pore-free

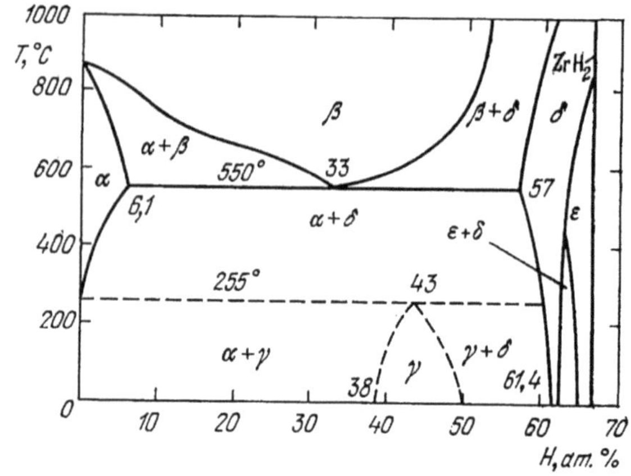

Fig. 4.30 Equilibrium diagram zirconium–hydrogen

material, and relative straightforwardness of the process. Among the disadvantages of this method are prolonged cycle time required for preform hydrogen impregnation (up to several weeks for large-sized performs), and the necessity to use autoclave equipment.

Forming the parts from hydride powders offers higher efficiency production of hydride parts and the capacity for composite production. The main disadvantage of the powder-based materials is typically imperfect grain boundaries that lead to poorer corrosion resistance in gaseous media. Reaction of Zr metal with hydrogen proceeds spontaneously and is accompanied by heat generation. Such spontaneous nature of reaction before hydride formation typically leads to cracking of the impregnated metal, caused by stress generation and simultaneous embrittlement. Cracking could be prevented through controlling the impregnation speed that leads to more balanced processes of stress generation and relaxation. The qualitative process and phase formation kinetics description could be obtained from the equilibrium phase diagram of the metal-hydrogen system. e.g. ZrH diagram (see Fig. 4.30).

The process may be approximately divided into three stages. First stage involves formation of a β solid solution of hydrogen in zirconium (above the temperature of polymorphic transformation). The second stage involves formation of a hydride phase (at first on a surface. then in the bulk). This stage brings bulk volume changes in the materials that generate stresses possibly exceeding the ultimate strength of the material. The third stage involves compositional homogenization of the hydride phase, accompanied by the redistribution of the stresses (both in the magnitude and in the sign). Reach-through Zr impregnation by H is accompanied by volume change. For hydrides with composition close to MeH_2 (e.g. for Zr hydride) this volume change may reach up to 20 %. The magnitude of the volume change influences the stress state level in the material being hydrogenated.

Based upon the data accumulated during research into hydrogen diffusion and creep [37, 38] Research Institute "Luch" has offered an optimized hydrogenation technology.

Table 4.7 Influence of temperature on strength σ B. thermal stress resistance ($\bar{R}$,R′) and physical properties $ZrH_{1.85}$ [6]

T (°C)	σ_B (MPa)	E·10^4 (MPa)	σ/E 10^3	α·10^{-6} (1/K)	λ (W/mk)	$\bar{R}$ (K)	R′ · 10^2 (W/m)
−190	-	-	-	4.0	-	85	-
20	29.4	6.86	0.4	6.0	28	70	20
200	40.2	6.08	0.7	0.8	32	70	22
400	53.9	5.29	1.0	12.8	34	70	24

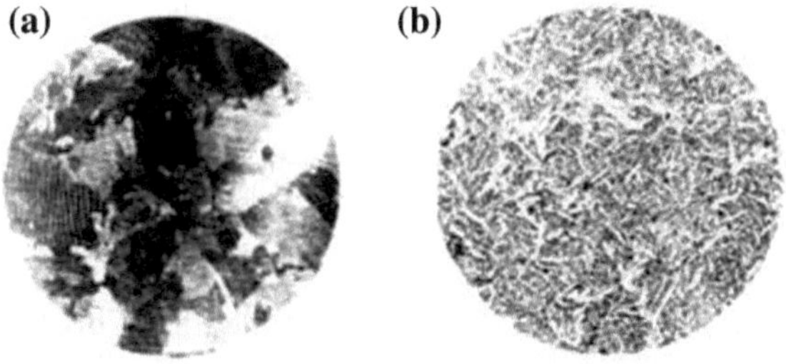

Fig. 4.31 A microscopic structure (x150) of not alloyed coarse-grained $ZrH_{1.85}$ with the martensitic structure ×150 (**a**) and fine-grained composition TiH with additives of Ti metal ×150 (**b**)

As the zirconium hydrides have insufficient strength and fracture toughness [36], there is a risk of their brittle fracture under thermal stresses generated under operating conditions. This was the driving force for research into the basic mechanical parameters of zirconium hydride. The thermal stress resistance of zirconium hydride samples processed by reach-through impregnation is characterized by highly specific temperature dependence (see Table 4.7).

Thermal stress resistance (TSR) of the samples is practically stable within temperature range of −190 to 900 °C despite the twofold strength disparity. The stability of experimentally measured TSR values is confirmed by the calculated R criterion, as the strength increase and E-modulus decrease temperature rise are compensated by significant growth of a linear expansion coefficient. The bulk structure of a sample and surface quality of hydrides after various machining treatments, influence both strength and TSR [36]. Samples after surface polishing demonstrate the largest TSR values.

The formation of hydride composites at introduction of a metal phase to 20–40 vol% reduces martensitic grain nature (Fig. 4.31) and increases thermal stress resistance by two to three times, due to a local stresses relaxation, increase of a fracture toughness and the relationship σ/E (Table 4.8).

4.6 Possible Methods for Increasing the Strength Parameters of Ceramics

The choice of technologies and parameters for manufacturing materials for improving the properties of ceramics with a limited plasticity in a broad temperature range should be performed taking its operation parameters into account. For materials of

Table 4.8 Modification of thermal stress resistance and mechanical characteristics of hydride composites with metal inclusions

Content	Inclusion size, $l/t\,(\mu)$	σ_b (MPa)	$\sigma/E\,10^{-3}$	K_{IC} (MPA) $M^{1/2}$	R (K)
$ZrH_{1.8}$, $d_3 = 500\,\mu$	-	30	0.43	1.7	80
$ZrH_{1.6}$ + 27vol%Zr	120/200	140	1.9	2.2	120
$ZrH_{1.7}$ + 19vol%Al	150/300	130	1.9	3.0	160
$ZrH_{1.5}$ + 30vol%Al	130/100	136	1.9	4.0	190
$ZrH_{1.7}$ + 22vol%Be	70/40	450	4.7	2.5	95
$YH_{1.9}$, $d_3 = 250\,\mu$	-	20	0.15	1.5	21
$YH_{1.1}$ + 40vol%Y	100/110	45	5.6	4.0	40
$TiH_{1.6}$, $d_3 = 20\,\mu$	-	150	1.3	2.2	75
$TiH_{1.2}$+18vol%Ti	150/3	390	3.5	3.4	140
$TiH_{1.8}$+40vol%Ti	30/5	840	7.6	7.6	300

l, t are length and width of particles. d_3 is aggregate size.

the first sections of the HRA operating in the brittle-damage temperature range, it is necessary to increase, along with the strength, the fracture toughness, which can be done first of all by decreasing the defectiveness of the materials and producing a structural state that increases the amount of energy required to produce damage. The strength in this temperature region is increased by the following methods: elimination of structural defects by optimizing the condensation and sintering regimes [15, 16], healing of defects and thermomechanical programmed control [21], modification of a stressed surface state [39], or carbon doping [17, 18, 27], or metallic band [14]. Essential increase of strength can be reached at use of nanocrystalline technologies [40, 41].

The elimination of surface defects by healing at temperatures $T/T_m > 0.5$ can increase the strength appreciably. The healing of radial cracks to the depth up to half a radius of the ZrC sample begins when the crack edges contact by surface diffusion. The healing is intensified with increasing the number of contacts on one length unit and is decreased with crack edge opening δ. At $T/T_m = 0.6$ the healing of the cracks with $\delta = 1{-}2\,\mu$ is completed within several hours and strength returns to the initial level (Fig. 4.32). The cracks with $\delta > 3\,\mu$ and very small number of edges contacts are not healed within the same length of time. Under thermal treatment conditions the healing process has two stages.

The kinetics of the first stage (the increased intensity of the strength reduction process) is mainly defined by surface self-diffusion of Zr atoms in ZrC. After the first stage of healing, the crack is a system of isolated cavities. Its healing is performed at a lower rate (the second stage) owing to a viscous flow of the material as the surface diffusion cannot provide their volume decrease.

Another way of ceramic materials strengthening is the formation (in surface layers) of compressive residual stresses preventing the appearance and spreading of cracks. The formation of residual stresses on the surface of a sample or an article is based on relaxation of thermo-elastic stresses nonuniformly distributed along the section.

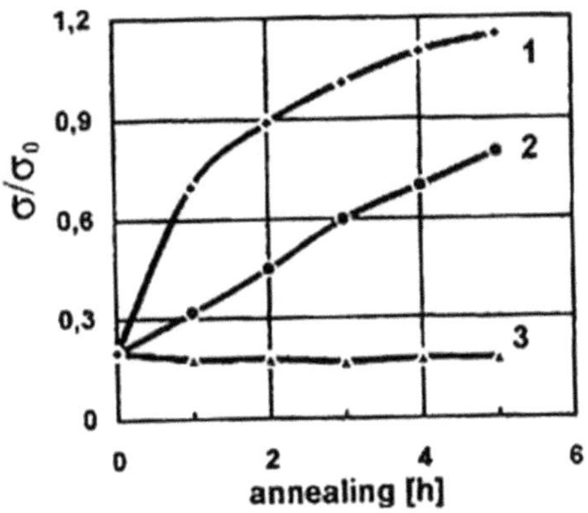

Fig. 4.32 The strength reduction of $ZrC_{0.95}$ samples after healing of the surface cracks with the width of crack edge opening (*1*—I μ, *2*—1.5 μ, *3*—3–4 μ) at $T = 2{,}800\,K$

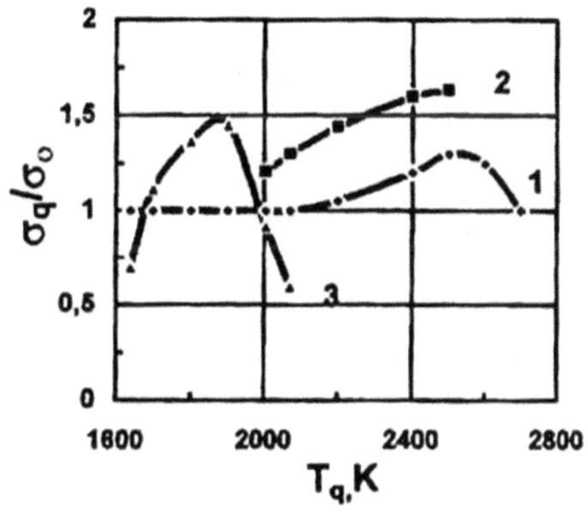

Fig. 4.33 The bending strength change of $ZrC_{0.95}$ (*1, 2*) and Sc_2O_3 (*3*) in relation to quenching temperature (T_q) and cooling methods. *1* radiation cooling; *2* gaseous helium flow; *3* cooling in silicon oil

As a result of poor ductility of ceramic materials, in contrast to metals, their strengthening is performed at rather small values of criteria Bio$\sim 10^{-1}$ e.g. by blowing off a cold gas stream over the heated sample or by radiation cooling [6]. At higher Bio values the rate of thermal-elastic stress relaxation turns to be less than the rate of their increase, this leads to cracking. The temperature range of strengthening is limited in the lower domain by the temperature of the brittle-ductile transition and in the upper domain by causing the strength decrease. The increase of temperature in the range $T_{b-d} < T < T_s$ rises the strengthening effectiveness. The gain in strength of strengthened ceramics is, as a rule, 20–40 % (Fig. 4.31).

A decrease of the critical volume defects, as stress concentrators, is possible by using thermal-mechanical treatment (TMT) at $T/T_m > 0.6$, based on stress relaxation near the concentrators [21].

The preliminary small deformation ($\varepsilon < 0.15\,\%$) at the low deformation rate $\varepsilon' < 10^{-1}\,s^{-1}$ (Fig. 4.34a) or static loading at stresses up to 0.6–0.8 σ/σ_{max} (Fig. 4.34b) can increase the strength about two times. As said above there is a wide

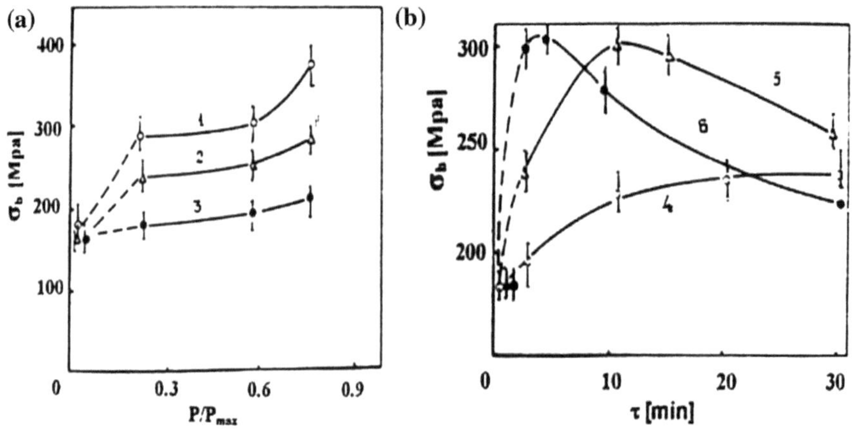

Fig. 4.34 The zirconium carbide strength change ε at T = 280 K after the high temperature thermal mechanical treatment at (T = 2100 K ε up to 0.15 %) with the sample loaded at different deformation rates **a** (*1*) $\varepsilon' = 10^{-4}\,s^{-1}$; (*2*) $\varepsilon' = 10^{-3}$; (*3*) $\varepsilon' = 10^{-2}$ and static loading; **b** at T = 2,100 K, (*4*) $\sigma/\sigma_{\max} = 0.3$; (*5*) $\sigma/\sigma_{\max} = 0.6$; (*6*) $\sigma/\sigma_{\max} = 0.8$

spectrum of volume defects in materials prepared by powder metallurgy methods. It is possible to evaluate the integral structural defects by measuring the value of temporary microstresses appearing under loading in the elastic region by broadening of X-ray lines and rejecting the defective samples. The more intensive is the line broadening with growth of load, the more defective is the material. In the temperature region where the macro-plastic deformation becomes possible, the short- and long-term mechanical thermal-strength parameters can be optimized by almost all methods used for metals. In the temperature region where the deformation process is controlled by the motion of dislocations, the methods are used that reduce the dislocation mobility by substructural strengthening, doping of solid solutions resulting in the formation of stronger chemical bonds in compounds, and doping with the formation of second phases. Under loading conditions, when deformation is mainly caused by grain-boundary sliding in the case of high-temperature creep, the strength can be efficiently increased by recrystallization, providing a considerable decrease in the length of boundaries due to the increase in the grain size. In using strengthening methods, it is necessary to take into account that structural changes intended to improve mechanical characteristics at high temperatures should not impair mechanical properties at temperatures $T < T_{b-d}$. The optimal choice of the parameters of traditional isothermal sintering of ZrC samples in different media (hydrogen. argon. and vacuum) allows achieving a high density (no less than 95 %) and the bending strength about 550 MPa at sintering temperatures 2,500–2,700 K [16].

References

1. Jackson, H. F., & Le, W. E. (2012). Properties and characteristics of ZrC (pp. 339–372). Amsterdam: Elsiver.
2. Andrievsky, R. A., Spivak, I. I. (1989). Strength of refractory compounds; Directory (p. 367). Chelyabinsk: Metallurgy.
3. Andrievsky, R. A., Lanin, A. G., & Rymashevsky, G. A. (1974). Strength of refractory compounds-M: Metallurgy, p. 232.
4. Daragan, I. D., D'jakov, E. K., Fedik I. I. et al. (2003). Fuel element assemblages of the space nuclear power propulsion systems. Moscow: Nuclear Technology Engineering Industry. vol. IV-25, under Adamov's edition. Engineering industry, book 2.
5. Nuclear Technology Engineering Industry (2003). Moscow, vol. IV-25, under Adamovs's edition. Engineering industry., book 1. 953 p. book 2. 943 p.
6. Lanin, A. G., & Fedik., I. I. (2008). *Thermal stress resistance of materials* (p. 239). Heidelberg: Springer.
7. RIPRA "Luch". (2004). Affairs and people. Podolsk. RIPRA "LUCH". Ed. Fedik I.I. Podolsk, 455 p.
8. Andrievsky, R. A., & Umanskiy, Y. S. (1977). Interstitial phases M. Science, p. 238.
9. Lanin, A. G., Zubarev, P. V., & Vlasov, K. P. (1993). Research of mechanical and heat-physical properties of fuel and constructive materials of NRER. *Atomic Energy,74*(1), 42–47.
10. Lanin, A. G., & Babajants, G. I. (2003). Structural materials of an active zone of NRER. In *Engineering nuclear industry* (vol. IV-25). under E.O.Adam's edition. Moscow: Engineering Industry. book 1.
11. Kosycheva, L. I., Lanin, A. G., Manjuhin, V. P. et al. (1978). Research of physical-mechanical properties of fuel compositions ZrC-UC. ZrC-NbC-UC. ZrC-UC-C (vol. 111803, pp. 47–67). Podolsk: Scientific Research Institute NPO "LUCH".
12. Andrievsky R. A. (1991). *Powder materials technology* (p. 207). Moscow: M. Metallurgija.
13. Andrievsky, R. A., Hromonozhkin, V. V., et al. (1969). Evaporation of uranium carbide. uranium nitride and uranium carbide-nitride. *Atomic Energy,26*, 494–497.
14. Lanin, A. G. (1998). *Strength and thermal stress resistance of structural ceramic* (111 p.). Moscow: M. Moscow State Engineer Physical institute.
15. Nezhevenko, L. B., Groshev V. I., & Bokov, I. V. (1970). Influence of ZrC powders on properties of sintering samples. In "Refractory Carbides" (pp. 58–61). K. Naukova Dumka.
16. Bulychyov, V. P., Andrievsky, R. A., & Nezhevenko, L. B. (1977). Sintering of zirconium carbide. *Powder Metallurgy,1*(4), 38–42.
17. Gerasimov, P. V., Egorov, V. S., Lanin, A. G., Nezhevenko, L. B., & Sokolov, V. A. (1982). Strength of compositions on the basis of zirconium carbide with disperse carbon inclusions. *Powder Metallurgy,11*, 67–74.
18. Lanin, A. G., Popov, V. I., Maskaev, A., et al. (1981). Strength of carbide-graphite compositions at power and thermal loading. *Problems of Strength,112*, 89–95.
19. Lanin, A. G., Marchev, E. V., & Pritchin, S. A. (1991). Non-isothermal sintering parameters and their influence on the structure and properties of zirconium carbide. *Ceramics International,17*, 301–307.
20. Lanin, A. G. (2007). Physical processes microindentation of carbide monocrystals of transition metals. *Functional Materials,1*(110), 383–389.
21. Lanin, A. G. (2004). Strength and thermo-mechanical reinforcement of the refractory ceramic materials. *Izvestia of the Russian Academy of Sciences. Series Physical, 68*(110), 1503–1509.
22. Zubarev, P. V. (1985). *Heat resistance of interstitial phases* (101 p.). Moscow: M. Metallurgija.
23. Derjavko, I. I., Egorov, V. S., Lanin, A. G., et al. (2001). Radiographic research of the residual Stresses in the rod carbide fuel elements. *The Bulletin of the National Nuclear Centre of Republic Kazakhstan,14*, 95.
24. Fedik, I. I. Kolesov, V. S., & Mihajlov, V. N. (1985). Temperature fields and thermal stresses in nuclear-reactors (280 p.). Moscow: M. Energoatomizdat.

25. Lanin, A. G., Erin, O. N., & Turchin V. N. (1990). Zirconium carbide strength and plasticity. *Refractory Metals and Hard Metals, 92*(3), 120–124; 139–141.
26. Zubarev, P. V., Kuraev, A. B., Lanin, A. G., et al. (1992). Influence of a grain size on creep of zirconium carbide. *FMM,16*, 122–125.
27. Lanin, A. G. (1995). Thermal stress resistance of porous Si_3N_4. ZrC heterogeneous carbides and hydrides. *Proceedings of 6th International Symposium on Fracture Mechanics of Ceramic, July 18–20, Karlsruhe, Germany*.
28. Lanin, A. G., & Egorov, V. S. (1999). Elastic–plastic fracture of the bodies under combined influence of the thermal and mechanical loads. *FCHOM,2*, 78–81.
29. Vlasov, N. M., & Fedik, I. I. (2001). *Fuel elements of nuclear rocket engines* (p. 207). Moscow: Tsniatominform.
30. Zelenskij, D. I., Pivovarov, O. C., Tuhvatulin, S. T. et al. (1999). Experience generalization of reactor working off of the rod carbide fuels on a complex stand "Baikal-1" and the development of applied production engineering (pp. 49–60). *The Fifth International Conference "Nuclear Power Space", Podolsk*.
31. Katz, S. M. (1981). *High-temperature heat insulating materials* (p. 232). Moscow: M. Metallurgy.
32. Vjatkin, S. E., & Deev, A. N. (1967). *Nuclear graphite* (279 p.). Moscow: Atomizdat.
33. Kats, S. M., Gorin, A. I., & Semenov, M. V. (1972). *Poroshkovaja Metallurgija,7*, 87–92.
34. Babajants, G. I., Golomazov, V. M., Granov, V. I., & Shmakov, V. A. (1974). Copyright certificate N 424658 "Opening. inventions. commercial machines. Trade marks", vol. 15, p. 38.
35. Ponomarev-Stepnoy, N. N. (1993). Creation history of NRER in the USSR (pp. 3–18). *Third Branch Conferences "Nuclear Power in Space"*.
36. Andrievsky, R. A. (1986). *Material science of hydrides* (p. 129). Moscow: M Metallurgy.
37. Lanin, A. G., Zalivin, I. M., & Turchin, V. N. (1984). Mechanical property of hydride alloys Zr. Y. Ti. *Problem of Strength, 6*, 83–86.
38. Zubarev, P. V., & Ryzhov, P. (1979). Creep of Zr and Y hydrides inorganic materials. 16(2), 247–250.
39. Lanin, A. G. (2011). Influence of residual stresses on ceramic materials strength and fracture. *Deformation and Fracture, 4*.
40. Koch, C. C., Ovud'ko, I. A., Seal, S., & Veprek, S. (2007). *Structural nanocrystalline materials: Fundamental and application* (p. 364). Cambridge: Cambridge University Press.
41. Andrievsky, R. A. (2012). *The basis of nanostructural materialscience: Possibilities and problems* (p. 251). Moscow: Publishing house BINOM.

Chapter 5
Radiation Resistance of the HRA Elements

The radiation estimation of fuel elements (FA) and structural materials resistance to irradiation damage implemented during tests in reactor IVG-1 [1] and at measurements of properties after irradiation. The neutron flux and test temperatures in special reactor loops varied within 10^{12}–10^{15} n/sm^{2}s and 450–2,000 K accordingly [2, 3]. The uranium burnout in FA at propulsion mode (PM) of NRER within 1 h attained $\sim 5 \cdot 10^{15}$ and $\sim 2 \cdot 10^{20}$ fiss/sm^3 at a power regime during 5,000 h. Resource tests were carried out at heat release levels 15–35 kW/sm^3 during 300–4,000 s. 200 FA had been tested in all.

The general regularity of materials' behavior under irradiation may be regarded in the monography [4] and some information on interstitial phases there is in [5, 6], but the data on irradiation changes of fuel materials properties of the solid solutions with uranium of carbides and nitrides of 4–5 groups of periodic table are absent.

The most important characteristic of the radiation resistance of HREs is their dimensional stability. The swelling of HREs made of compositions based on solid solutions of uranium, zirconium, and niobium carbides depends on the fission density and irradiation temperature in a reactor is nonmonotonic function [3]. An increase in the irradiation dose of UC + ZrC + NbC and UC + ZrC fuel compositions up to 2×10^{19} fissions per cm^3 at the irradiation temperature $T = 1{,}100$ K leads to swelling by 5 % and an increase in the electric resistance up to 80 % due to the accumulation of radiation defects (mainly vacancies) (Fig. 5.1).

The bubble swelling change of fuel volume during an irradiation incubatory period is defined by balance between two competitive processes: swelling at the expense of accumulation of radiation-induced defects and irradiation sintering [3]. The swelling during the incubatory period poorly depends on the type of the investigated compositions and is defined, substantially, by an irradiation temperature. As concentration of the fission products is still the lowest, the dimensional sample modification is caused by accumulations of the dot defects, which evolution leads to gradual swelling growth under irradiation. Calculation shows that the maximum contribution of solid swelling at the expense of the accumulation of fission products, at $B = 2 \cdot 10^{20}$ fiss/sm^3 makes 0.85 %., i.e., about 35 % from the total. Thus, dimensional changes of the investigated

A. Lanin, *Nuclear Rocket Engine Reactor*, Springer Series in Materials Science 170,
DOI: 10.1007/978-3-642-32430-7_5, © Springer-Verlag Berlin Heidelberg 2013

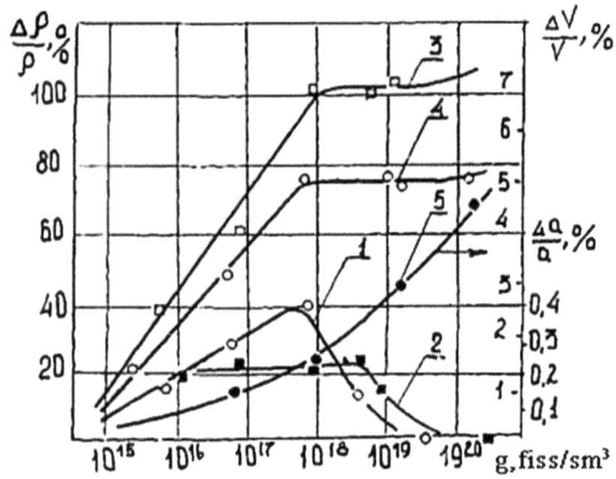

Fig. 5.1 Change of the lattice spacing $\Delta a/a$ (*1, 2*), specific electrical resistance $\Delta\rho/\rho$ (*3, 4*) and swelling ΔV/V (*5*) of fuel materials ZrC of +8 % UC (P = 5 % porosity) (*1, 3, 5*) and ZrC+8 mass. % UC +1 mass. % C (*2, 4*) depending on fission-density at irradiation temperature T = 1,100 K

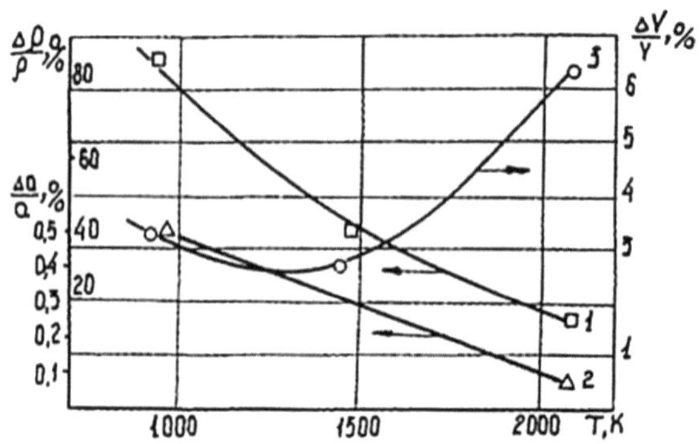

Fig. 5.2 Change of specific electrical resistance Δρ / ρ (*1*), lattice space ratio Δa/a (*2*) and volume ΔV/V (*3*) of fuel material ZrC +8 mass. % UC (P = 5 %) in dependence of irradiation temperature and fission density $2{\cdot}10^{20}$ f/cm^3 [7]

compositions irradiated at T = 1,000 K and B from $8.4{\cdot}10^{17}$ to $2{\cdot}10^{20}$ fiss/sm^3 are defined by the swelling at the expense of the accumulations of radiation-induced defects (predominantly vacancies).

Gas bubble swelling of the fuel compositions UC + ZrC + NbC, UC + ZrC at temperature of irradiation T = 2,100 K, begins at a burnout A = $2{\cdot}10^{19}$ fiss/cm^3. The volume modification $\Delta V/V_0$ from initial sample porosity p_0 irradiated in the limits from $7{\cdot}10^{17}$ to $1.8{\cdot}10^{19}$ is defined by the constitutive equation:

$$\Delta V/V_0 = -C_0 P_0[1 - \exp(-B/B_0)],$$

where C_0, B—the constants depending basically from temperature irradiation and type of a fuel composition, P_0 = initial porosity, B is a burnout.

The rise of irradiation temperature of the same fuel compositions from 1,000 to 2,100 K at constant fission density $2{\cdot}10^{20}$ fiss/sm^3 increases swelling to 6 %, but reduces the values of lattice constant Δa/a and electrical resistance Δρ/ρ to the initial value without irradiation [8] that is connected with sufficiently high annealing rate of radiation-induced defects (Fig. 5.2).

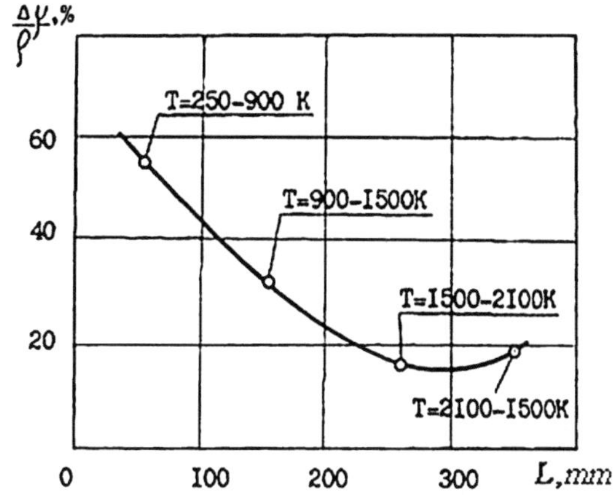

Fig. 5.3 Relative drops of the fuel element resistance $\Delta\rho/\rho$ (%) along wise of the central HGA with change of the fuel element temperature

Noticeable changes of the fuel element electrical resistance $\Delta\rho/\rho$ are observed in the central HGA with temperature variation of the fuel elements (Fig. 5.3).

The microscopic structure is changed simultaneously with swelling. The growing pores are arranged mainly on the grain boundaries and graphite inclusions. The most increment of porosity about 15–20 %, is revealed for 3-fold solid solution UC-ZrC-NbC. Origination of pores increases the specific electrical resistance and decreases modulus of elasticity that proves to be true to the ρ and E calculations on porosity at known relationships [2].

The research of the carbide fuels, implemented on the basis of the developed express complex techniques [1], confirms the radial fuel inhomogeneity swelling effect on a magnitude and on a stress sign of the fuel surfaces as at high-and low-temperature sections. Dependence of lattice constant change $a_0 = f(R)$ along fuel radius is close to the parabolic law [8, 9] (measurements of a_0 were carried out by X-ray survey on various sections of a strongly oblique fuel section) (Fig. 5.4).

Different concentrations of radiation defects along the HRE radius (higher on the surface and lower at the more heated center) due to annihilation of point defects lead to a parabolic change in the lattice space. This gives rise to compressing stresses in the surface layers resulting in an increase of the strength and thermal strength. UC-ZrC and UC-ZrC-NbC HREs irradiated by the neutron flux $J = 10^{14} - 10^{15}/cm^2/s$ at the fission density $B = 10^{16} - 10^{18}/cm^3$ and $T < 0.4T_m$ enhances the strength by 30–50 % and the thermal strength resistance by 70–80 %. In this case, the strength increment decreases upon increasing the fission density above 10^{18} fissions per cm^3 (Fig. 5.5) due to the formation of vacancy and gas pores and some amorphization of materials.

The strength loss of carbide graphite (UC-ZrC+5 mas. %) occurs at lower burnout in comparison with double and three-fold solutions. The double solution UC-ZrC at irradiation is stable to fission density not less than $2{\cdot}10^{19}$ fiss/cm^3.

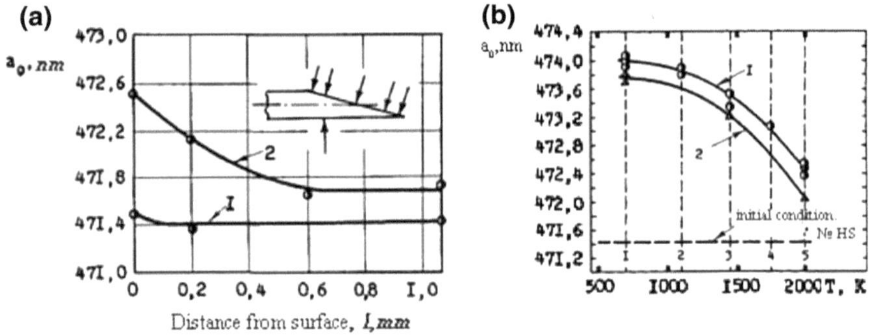

Fig. 5.4 Space lattice change on section of twisted fuel elements 2.2 mm in diameter in fifth heating section (HS-5) before (*1*) and after (*2*) irradiation (**a**); degree of fuel element swelling, estimated by the surface lattice space (*1*) and by medium body lattice space (*2*) for fuel elements of the different HS (**b**)

Fig. 5.5 Dependences of changes in the HRE strength on the fission density at temperatures 800–1,000 K: (*1*) UC—ZrC-NbC; (*2*) UC-ZrC; (*3*) UC-ZrC+5 %C [9]

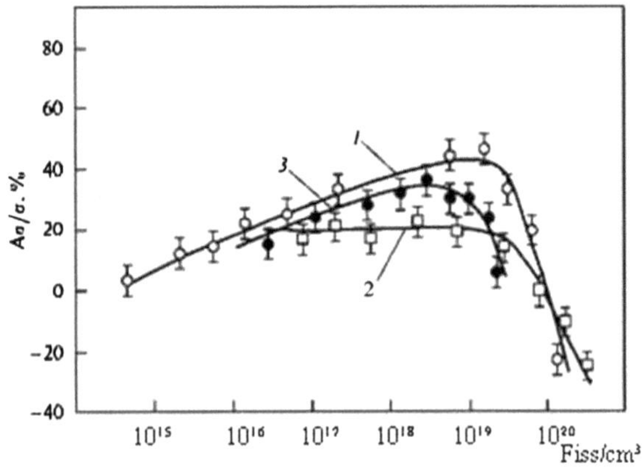

Dependence of residual stresses σ_{rs} acting on a surface at inhomogeneity swelling [8]: can be easily computed by the change of lattice constant a_0^s on a fuel surface and measured modulus elasticity value E

$$\sigma_{rs} = [A/(1-\nu)] \times [(\Delta V/V)] = [E/(1-\nu)] \times \left[\left(a_0^s - \overline{a_0}\right)\right]$$

The isochronous annealing of heated HREs at 1,500 K removes inhomogeneity of swelling and the residual stresses, returning the strength to its initial level before irradiation, whereas the same irradiation doses in the case of structural carbides cause only weak changes in the elastic modulus and strength, along with a noticeable increase in the lattice pitch and electric resistance (Fig. 5.6).

Irradiation of structural carbide materials under conditions similar to those of irradiation of fuel materials, while retaining the volume and shape of the samples, leads

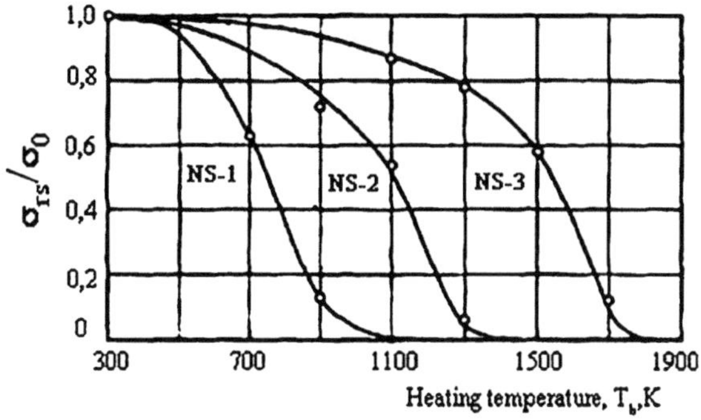

Fig. 5.6 Residual stress change σ_{rs} / σ_0 after fuel elements isochrones annealing of three HS of one HGA

to an appreciable growth of the lattice parameter and growing electrical resistance, accompanied by a minor change of strength and Young modulus [10].

Under the parameters being studied, the Zr carbide density increased, while that of Nb carbide shows virtually no significant change. Lattice parameter and electrical resistance of ZrC increase much more significantly that those of NbC, while microhardness of these materials shows inverse behavior. After irradiation, the cracking resistance (as measured by change of the load P on the hardness tester' indenter) of ZrC drops much more than that of NbC, while their thermal stability grows approximately by the same factor 1.7. Equimolar solid solution of ZrC and NbC behaves similarly to ZrC. though ρ behavior is rather close to that of NbC, while thermal stability demonstrating its own, unique behavior.

It is important to note that an absolute gain of electrical resistance after irradiation is much higher in ZrC and NbC, as compared to that in irradiated metals, which confirms the significant effect of the carbon sublattice on formation of the irradiation-induced defects as measured by electrical resistance change. Difference in ZrC and NbC behavior under irradiation should be attributable to the general differences in physical and chemical properties of IV and V group carbides, due to the electronic structure features [10] (Table 5.1).

It is also typical that, with increasing irradiation temperature to 1,300 K, the difference in ZrC and NbC behavior remains virtually the same as at 450 K, though radiation damage consequences become much smaller. Particularly, noteworthy is also the fact of increased thermal stress resistance, possibly attributable to generation of residual compressive stresses and to the effect of radiation-induced healing.

Radiation-induced fuel rod healing phenomenon has been experimentally confirmed by declining effect of the introduced surface cracks on UC-ZrC and UC-ZrC-NbC fuel rods after low-power irradiation inside the He-filled ampoules. The degree of crack healing was estimated by monitoring changes in strength σ. Electrical resistance ρ and an elastic flexure *f* of the fuel rods in the middle under consolidated mass impact.

The results of tests show that purely thermal annealing at relatively low (for carbide) temperature of 1,100 K cannot lead even to partial healing of the surface

Table 5.1 Physical-mechanical properties of irradiated samples

Properties	Materials		ZrC	NbC	(Zr, Nb)C
γ (g/cm)^2	Initial		6.40	7.4	6.97
	Irradiation	150 °C	6.29	7.4	6.84
		1,100 °C	6.27	7.4	6.91
a (Å)	Initial		4.692	4.471	4.575
	Irradiation	150 °C	4.714	4.488	4.598
		1,100 °C	4.698	4.472	4.578
ρ ($\mu\Omega \cdot cm$)	Initial		43	50	68
	Irradiation	150 °C	250	90	100
		1,100 °C	65	50	68
$\frac{\sigma_{И}}{\sigma_{Иmin}-\sigma_{Иmax}}$ (kgf/mm^2)	Initial		$\frac{250}{220-310}$	$\frac{350}{300-410}$	$\frac{380}{320-400}$
	Irradiation	150 °C	$\frac{320}{240-450}$	$\frac{350}{310-300}$	$\frac{280}{260-340}$
		1,100 °C	$\frac{260}{240-300}$	$\frac{390}{340-410}$	$\frac{340}{310-490}$
E $\cdot 10^{-3}$ (kgf/mm)^2	Initial		41	49	46
	Irradiation	150 °C	41.5	47.5	45.5
		1,100 °C	41	50	47
H_{μ} (kg/mm^2)	Initial		2,050	1,400	1,600
	Irradiation	150 °C	2,300	2,500	1,900
		1,100 °C	2,200	1,600	2,000
P, g	Initial		100	120	120
	Irradiation 150 °C		40	100	60
$\overline{\Delta T}$(°C)	Initial		45	70	500
	Irradiation	150°C	75	130	30
		1,100°C	75	135	50

Table 5.2 Average relative values change of fuel elements characteristics (f. ρ .σ) of UC-ZrC-NbC; with created surface cracks after irradiation in the nuclear reactor

Property	Initial value	After introduce of cracks	After annealing T = 1100 K, τ = 290 h	After irradiation T = 1100 K, τ = 290 h	After additional irradiation at 1870 K, τ = 1.5 h
f	100	124	125	104	100
ρ	100	126	125	141	103
σ	100	45	47	198	207

cracks (see Table 5.2). On the other hand, complete crack healing was observed after irradiation T = 1,100 K or after additional irradiation.

It is particularly apparent that crack-induced halving of fuel rod strength upon irradiation was replaced by 98 % strength rebound, of which 38 % correspond to a irradiation reinforcement observed in crack-free fuel rods, and the balance 60 % are attributable to an additional structural defect healing in rod surface, due to some of the critical surface defects being consumed during crack generation. The observed accelerated defect healing under irradiation is probably attributable to ‘displacement peaks' having much smaller size than the thermal cracks. The strength gain did not disappear after additional annealing at temperature above T_{irr}, whereas the flexure and electrical resistance returned to their initial values (i.e., before irradiation and crack introduction).

It should also be noted that the results of irradiation researches on the stability of the thermoelectric temperature transducers, for the determination of the possible temperature measurement errors. It was shown [11] that reversible and irreversible changes of the thermal electromotive forces are manifested under act of reactor irradiations. The reversible changes caused by additional energy release in small volume of a thermojunction are negligible small, irreversible changes grow appreciably with an increase of fluence neutrons and errors components should be considered necessarily as the regular capacity. The regular component of an error measurement ΔT_r for thermocouple tungsten-rhenium VR5/20 is defined by an aspect of dependence [11]:

$$\Delta T_r = a_\varphi\, \varphi_T + a^2 \varphi \sigma_{eaj} \cdot T_{irr},$$

where φ_T, φ_b—fluences of thermal and fast neutrons and T_{irr} is an irradiation temperature.

5.1 Radiation Durability of Graphitic Materials

As the graphitic materials are used in the HIP, irradiation durability of some graphite was investigated, especially at high temperatures 1,400–1,700 K [12]. First of all, the shrinkage rate of graphite is the most interest. For many types of graphite (except the isotropic MPG-6 graphite based on noncalcined coke) at high irradiation temperatures from 1,400 to 1,700 K in the interval of the neutron fluxes under study (up to $2.75 \cdot 10^{25}\,m^{-2}$), the shrinkage of samples was observed in both the parallel and perpendicular directions of the cut with respect to the formation axis. A decrease in the irradiation temperature to 1,200–1,500 K reduces the shrinkage rate and the anisotropy in the change of geometrical dimensions (Fig. 5.7).

At the same time, the relative change in the specific electric resistance for all graphite decreases, while the relative change in the elastic modulus increases. Such a behavior of graphite materials can be explained by considerable damage to the microscopic structure and the appearance of pores and cracks, indicating 'secondary swelling.

5.2 Radiation Durability of Moderator Material

Radioactive changes in the moderator material, zirconium hydrate, during NRE operation in the engine regime (1,000 s) are insignificant. In the power regime after the irradiation of zirconium hydride in a reactor by the integral neutron flux of the order of

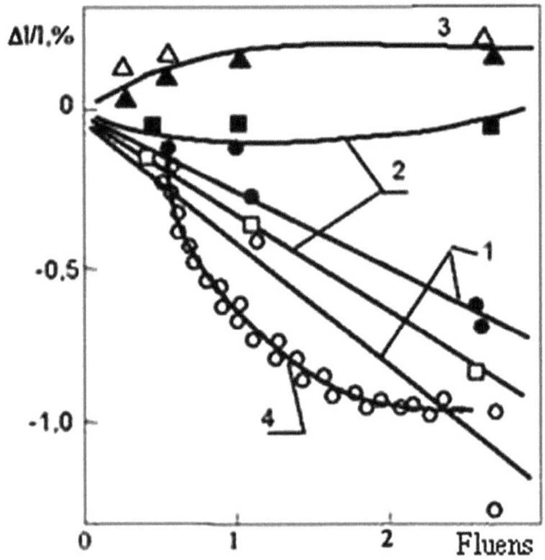

Fig. 5.7 Relative length change of graphite samples of various grade: RBMK (*1*), VPG (*2*), MPG 6 (*3*), irradiated under neutron fluence at temperatures 1,570–1,770 K. ■ ● ▲ Samples are cutting perpendicularly to compaction axis; □, ○, △ Samples are cutting in parallel direction to compaction axes; *1, 2, 3* measurements after irradiation; *4* measurements during irradiation

10^{21} cm^{-2}, hydride swelling by 1.5 % is possible [13]. The temperature dependence of the zirconium hydride swelling is nonmonotonic. The minimum of swelling is observed at 550 K and at 800 K is the maximum. The electric resistance changed on 7 % after 3,000 h of irradiation by a fast neutron flux with the density $3{\cdot}10^{20}$ cm^{-2} at 855 K [14]. The heat conduction of $ZrH_{1.8}$ irradiated by a fast neutron flux with the density $3.2{\cdot}10^{21}$ cm^{-2} at 320 K decreased from 31 to 18 W m·K^{-1} [14]. As a whole, the results of irradiative studies suggest that the efficiency of the HRA in the NRE in the engine and energy regimes is retained.

References

1. Derjavko, I. I., Perepelkin, I. G., Pivovarov, O. C., et al. (2001). Express techniques for research of the irradiated rod carbide fuels. *The Bulletin of the National Nuclear Centre of Republic Kazakhstan,4*, 88–93.
2. Chernetsov, M. V., Sernjaev, G. A., & Scherbaks, E. N. (1999). Influence of irradiation on sizes and physical-mechanical properties of the carbide fuel. Nuclear Power in Space. Nuclear Rocket Engines. Podolsk. part 2. p. 259.
3. BπacoB, H. M., & Zaznoba, V. A., (1999). The incubatory swelling period of fuel materials SIA "LUCH", Podolsk, Russia. Nuclear Power in Space. Podolsk. 1999. Reports. part 2, pp. 489–491.
4. Radiation damage of refractory compounds. (1979). Under M. S. Ogorodnikova. J.I. Rogovogo's edition. Atomizdat. 160 p. (in Russian).
5. Jackson, H. F. & Le, W. E. (2012). Properties and characteristics of ZrC (pp. 339–372). Amesterdam: Elsiver.
6. Was, G. S. (2007). *Fundamentals of radiation materials science. Metals and alloys* (p. 827). Heidelberg: Springer.

7. Derjavko, I. I., Zelenskij, D. I., Lanin, A. G. et al. (1999). Reactor working off of the rod carbide fuels of NRER (pp. 539–543). *5th International Conference "Nuclear Power in Space", Podolsk.*
8. Derjavko, I. I., Egorov, V. S., Lanin, A. G., et al. (2001). Radiographic research of the residual Stresses in the rod carbide fuel elements. *The Bulletin of the National Nuclear Centre of Republic Kazakhstan,14*, 95.
9. Deryavko, I. I., Chernyadev, V. V., Gorin, N. V., et al. (2010). Characteristics of identifying the fuel rods of NRE. *Vestnik NNC RK. edn*, *4*(44), 95–105.
10. Andrievsky, R. A., Vlasov, K. P., Lanin, A. G., et al. (1978). Influence of irradiation on physical-mechanical properties of zirconium and niobium. *Non-organic Materials,1*(4), 680–683.
11. Prijmak S.V., Olejnikov P.P., Taubin M. L., Features of metrological temperature measurement by thermoelectric temperature transducers in the conditions of intensive reactor irradiation. *The Third Branch Conference "Nuclear Power in Space" 1993* (pp. 327–333). *Atomic energy 67*(3), 221–222.
12. Vlasov, K. P., Kolesnikov, S. A., & Virgil'ev, J. M. (1993). Radiation durability of carbon materials (pp. 805–816). *3rd Branch Conference "Nuclear Power in Space. Nuclear Rocket Engines.", Podolsk.*
13. Pinchuk P.G., Bjakov H., Barabash AA. et al. Researches of radiation swelling in the system zirconium - hydrogen. Atomic energy. 1977. Vol. 42. 1 I, pp. 16–19.
14. Andrievsky, R. A., Ja, M. V, & Savin, V. I. (1978). Study of the defect nature of transition metals hydrides. *Inorganic Materials*, *14*(9), 1664.

Chapter 6
Corrosion of Materials in the Working Medium

An important factor determining the operation conditions of HRAs is the corrosion resistance of materials in hydrogen with methane added [1, 2].

The extent of carbides interacting with gases is defined by gas activity with carbide components and gas partial pressures in a mix [3–5]. The maximum activity is the oxygen, which interact even at low partial pressures (10^{-3}–10^{-1} vol%) harder above than hydrogen; therefore, the interaction with oxygen will be observed first. The mechanism of ZrC interacting with oxygen can be presented by following equations (Table 6.1). The detailed oxidation mechanism description of the NbC and solid solutions ZrC-NbC is complicated by two reasons: first, there is a row of the flying oxides in the system Nb-O which is sublimated at high temperatures; second, the oxidation of these carbides was studied in rather narrow temperature intervals.

Kinetic characteristics of oxidation of ZrC and NbC are resulted in Figs. 6.1 and 6.2.

Determination of kinetic regularity in an intermediate interval of temperatures 1000–1300 K is complicated because of an oxide film fall and an unknown surface of oxidation. The resulted data on oxidation kinetics refer to the materials practically with a stoichiometric composition. Decrease of C/Me to 0.9–0.8 in all carbide materials reduces reaction rate constant. High-temperature oxidation (2,100–2,500 K) in hydrogen ($H_2 + 0.07 \div 0.1\,\%O_2$) leads to decrease in strength of ZrC approximately two times for the first 1,000 s of interacting (Fig. 6.3). At higher temperatures (>2,600 K) in the specified gas mix ZrC samples are fractured after 500 s of interaction. A source of rupture is a formation of metal zirconium on reaction 5 (Table 6.1). Metallographic analysis of a fractured places after the sample exposure at $T = 2{,}700$ K showed the segregations of metal zirconium in an oxide film.

The interaction of carbides with hydrogen at high temperatures is accompanied by structural changes, the formation of hydrocarbons, and a decrease in the carbide stochiometric according to the reaction

$$\mathrm{MeC} + y/2\,\mathrm{H_2} = \mathrm{MeC} + \mathrm{C_xH_y}.$$

A. Lanin, *Nuclear Rocket Engine Reactor*, Springer Series in Materials Science 170,
DOI: 10.1007/978-3-642-32430-7_6, © Springer-Verlag Berlin Heidelberg 2013

Table 6.1 Change of ZrC oxidation mechanism with temperature

No	Reaction equations	Temperature range (°C)
1	$2ZrC + 3O_2 = 2ZrO_2 + C_{(\Gamma pa\Phi)} + CO_2$	<600
2	$ZrC + 2O_2 = ZrO_2 + CO_2$	600–1000
3	$ZrC + 3/2O_2 = ZrO_2 + CO$	1200–2000
4	$2ZrC + (x + y/2)O_2 = ZrCl - x.O_y + x\ CO$	2200–2500
5	$2ZrC + ZrO_2 = 3Zr + 2CO$	2400–2700

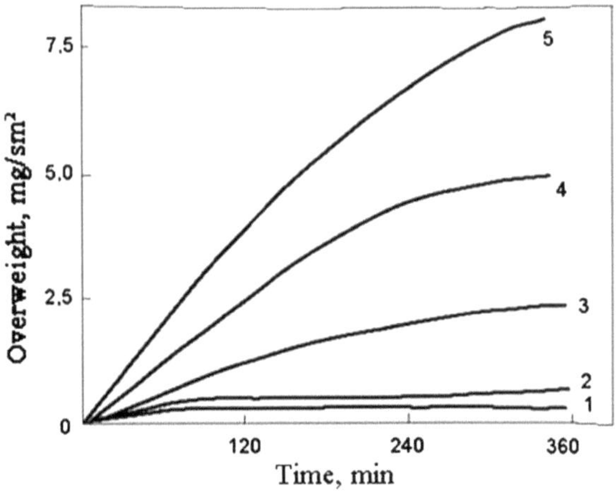

Fig. 6.1 Oxidation of zirconium carbide in air. ($ZrC_{0.95}$, $C_{total} = 11.4\,\%$. $C_{free} = 0.3$; $O + N = 0.10\,\%$; $\prod = 3{-}5\,\%$; $L = 8\,\mu$) [4]. *1*—T = 850 K, *2*—T = 900 K, *3*—T = 950 K, *4*—T = 1,000 K, *5*—T = 1,050 K

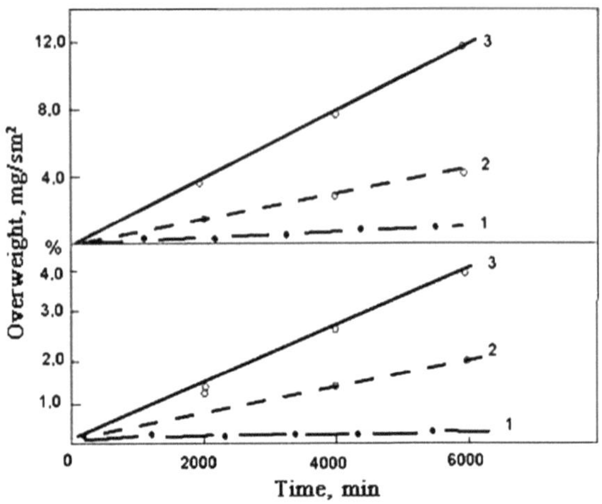

Fig. 6.2 Oxidation of zirconium carbide ($ZrC_{0.99}.C_{total} = 11.7\,\%$, $C_{free} = 0.2$; $O + N = 0.15\,\%$, $P = 3\,\%$, $L = 10\,\mu$) in technical hydrogen. Weight change (above); the oxygen maintenance in samples, mass % (below), *1*—T = 2,100 K, concentration O_2 in gas = 0.001 vol %; *2*—T = 2,500 K, concentration O_2 in gas = 0.1 vol %; *3*—T = 2,100 K, concentration O_2 in gas = 0.1 vol %

Change of lattice parameters of ZrC after an exposure time in hydrogen allows sizing up the change of relationship C/Zr on cross-section of the sample at the expense of a carbon removal (Fig. 6.4). Calculated estimations are coordinated qualitatively with the chemical analysis data.

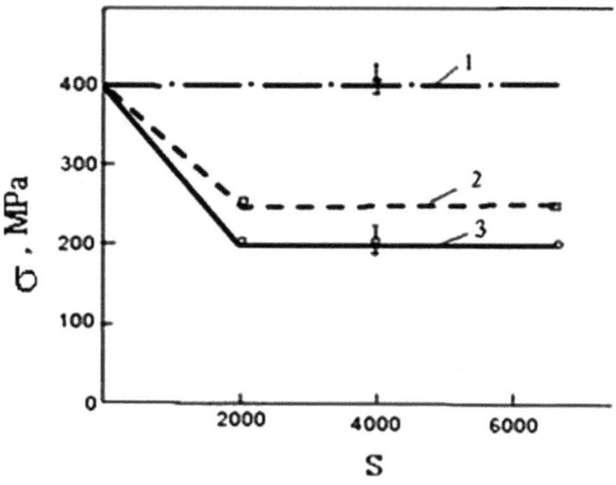

Fig. 6.3 The strength of zirconium carbide samples after oxidation in a technical hydrogen *1*— T = 2,100 K, concentration O_2 in gas = 0.001 vol %; *2*—T = 2,500 K, concentration O_2 in gas = 0.1 vol %; *3*—T = 2,100 K, concentration O_2 in gas = 0.1 vol %

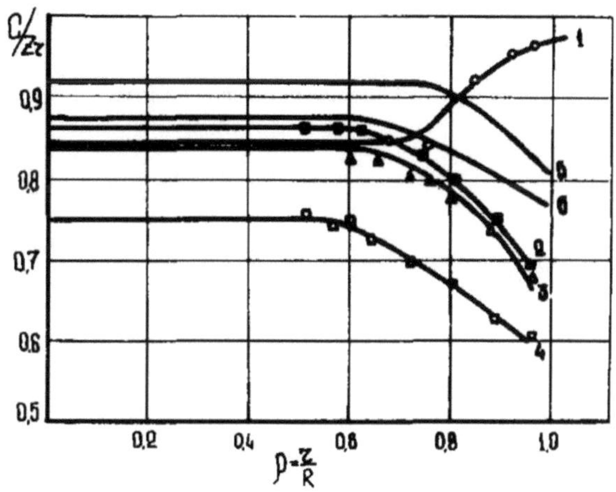

Fig. 6.4 Carbon content allocation on section of ZrC sample in an initial state-1 and under forced hydrogen convection at parameters T = 3,100 K, P = 2 MPa after exposure time; *2*—$\tau = 2 \cdot 10^3$ s; *3*—$\tau = 4 \cdot 10^3$ s. *4*—$\tau = 6 \cdot 10^3$ s.; calculated estimation: *5*—$\tau = 2 \cdot 10^3$ s; *6*—$\tau = 6 \cdot 10^3$ s

The type of the interaction of UC-ZrC, UC-NbC, and UC-ZrC-NbC systems with hydrogen is similar. Only the degrees of decarbonation, the carbon concentration gradients over a sample cross-section, and recrystallization temperatures are different. This suggests that all the systems under study have the same mechanism of interaction with hydrogen. On the surface of samples, the reaction of carbon with hydrogen proceeds with the formation of CH_4 and C_2H_2 and the evaporation of metal atoms, while the diffusion of carbon from the center to the surface and recrystallization occur in the bulk. Calculating the interaction for the scheme of the process under study requires solving nonstationary diffusion problems with the boundary conditions simulating a mass exchange with the working substance in the presence of heterogeneous chemical reactions (Fig. 6.5).

The total rate of a material removal for ZrC and NbC is maximum at first moments of an exposure time and the temperature dependence is expressed by the Eqs. (6.7) and (6.8):

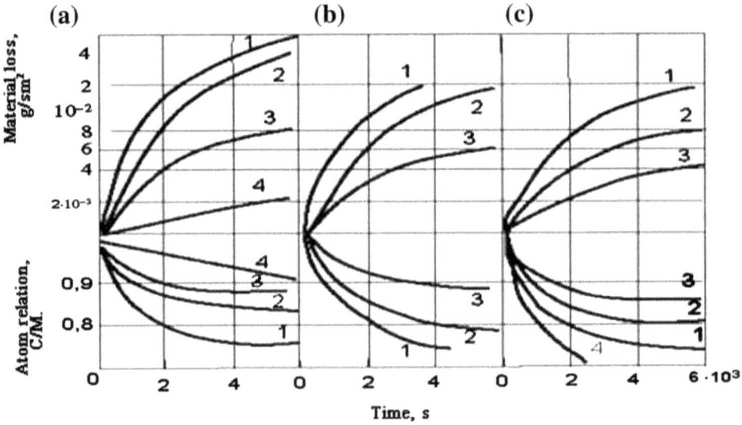

Fig. 6.5 Interaction of zirconium and niobium carbides, solid solution ($Zr_{0.5}$ $Nb_{0.5}$) C in hydrogen. **a** Zirconium carbide: *1* T – 3,400 K, *2* T – 3,100 K, *3* 1 – 3,000 K, *4* T – 2,600 K. **b** The solid solution *1*—T = 3,100 K, *2*—T – 3,000 K, *3*—T – 2,350 K. **c** Niobium carbide *1*—T = 2,980 K, *2*—T – 2,550 K, *3*—T – 2,580 K, *4*—T – 3,140 K

$$V_{ZrC} = 1.922 \exp(-0.612 \cdot 10^{-3} t) \exp\left(-\frac{87{,}000 - 3.89 \cdot 10^{-3} t}{RT}\right) \quad (6.7)$$

$$V_{NbC} = 0.058 \exp(-0.15 \cdot 10^{-3} t) \exp\left(-\frac{54{,}000 - 3.00 \cdot 10^{-3} t}{RT}\right) \quad (6.8)$$

The total entrainment rates for double ZrC-UC and triple ZrC-NbC-UC solid solutions in hydrogen–methane media with 0.656 volume % of CH_4 at 3,150 K are approximately the same ($0.94 \cdot 10^{-6}$ gcm^{-2} s^{-1}). High temperatures in experiments lead not only to a change in the chemical composition of materials but also to a change in the density of carbides and to an increase in the grain size. The appearance of a carbon concentration gradient over the sample cross-section at temperatures up to 2,500 K for 1,000 s leads to the formation of compressing stresses up to 500 MPa and to an increase in the strength due to a change in the lattice parameter.

The interaction of carbides with hydrogen at 3,100 K for 1,000 s produces an inhomogeneous concentration of carbon not only but also causes a change in the structure, a decrease in the density, and a reduction in the strength to 50 % (Fig. 6.6). It should be noted that exposure at a lower temperature near 2,600 K may even increase strength due to the formation of residual compressive stresses. The results of model tests agree well with the HRA tests in the IVG-1 reactor.

Decarburization of NbC in hydrogen occurs more intensive. Transformation of NbC into Nb_2C occurs at 2,450 K after 6,000 s. The gradient of concentration of carbon on cross-section is insignificant though Nb_2C grains are observed all over

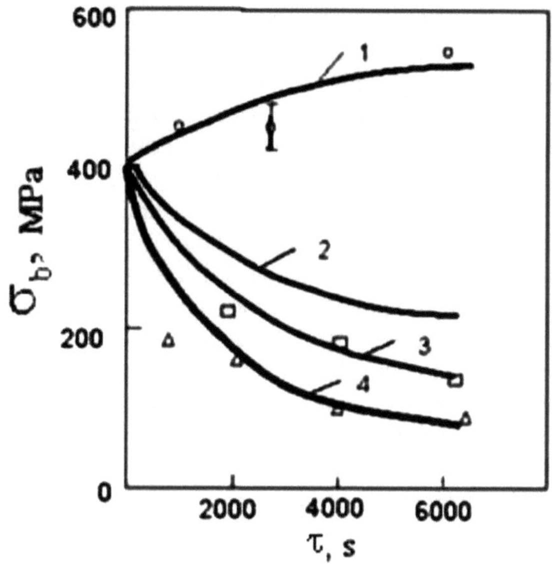

Fig. 6.6 Strength of zirconium carbide samples after an exposure time, τ in hydrogen *1*—T = 2,550 K; *2*—T = 3,040 K; *3*—T = 3,090 K; *4*—T = 3100 K

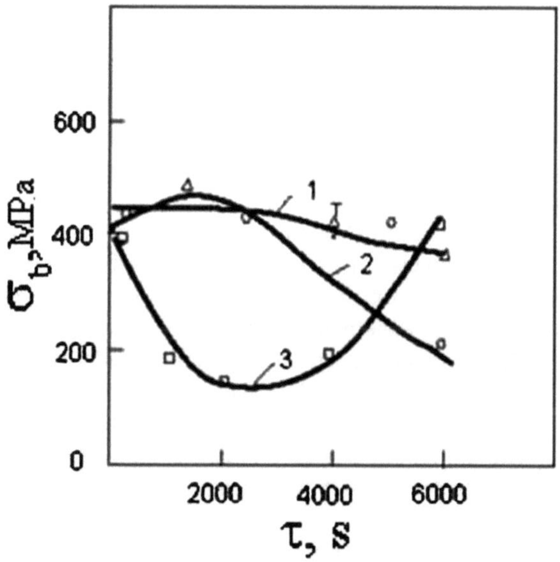

Fig. 6.7 The strength dependence of NbC samples after an exposure time in hydrogen is more complex than of ZrC. *1*—T = 2,550 K; *2*—T = 2,720 K; *3*—T = 2,950 K

cross-section. Active growth of grain is observed at a surface of the sample since 2,500 K. Temperature lifting to 3,000 K leads to intensive decarburization and the damage of samples (Fig. 6.7).

The general entrainment of niobium carbide is connected with carbon loss, and the metal component of carbide evaporates slightly, as opposed to ZrC when an appreciable evaporation of zirconium is observed. This fact supports higher concentration of carbon in ZrC after interacting. Interaction of solid solutions ZrC-NbC with hydrogen at high temperatures along with decarburization, the enrichment of surface by more refractory component of niobium is observed.

Apart from the chemical action of a high rate working substance flowing around an element, a force erosion action of the flow on the surface is also possible; as a result, some weakly attached carbide particles are carried away by the flow.

Based on the dependences of the entrainment rate of the carbon component from carbides, another protection method was proposed in which hydrocarbons are added to hydrogen in the amount at which their mean concentration in the flow becomes equal to their equilibrium concentration over the surface. Then the carbon entrainment must be zero. But the equilibrium composition of hydrogen can be achieved not over the entire HRA length but at individual points only. In other places, either the entrainment or deposition of carbon on the washed surface occurs. Therefore, it is reasonable to regard the method of protective additions as supplementary to the main method of protection with the help of coatings.

Thermal stresses, σ_1, in a coating of thickness h_1 originate even at uniform heating to temperature T due to a difference of coefficients of linear expansion of a coating α_1 and substrate α_2. Accepting the condition of equal deformation on the boundary line between coating and substrate, the stress in the coating can be expressed:

$$\sigma_1 = \frac{E_1 E_2 (\alpha_1 - \alpha_2) T}{\frac{h_1}{h_2} E_1 (1 - \mu_2) + E_2 (1 - \mu_1)}$$

where index 1 refers to coating and index 2 refers to substrate. E is modulus elasticity, μ is Poisson's coefficient. Resistance of coating can be determined by the level of limiting tensile stresses.

The thermal stress destruction of a coating can occur as a result of coating cracking or at the expense of its peeling from a substrate to the subsequent destruction of a brittle thin covering. The fracture aspect is defined by extent of adhesion and parameters of a thermal loading (Fig. 6.8).

The corrosion resistance of coating made of low-density pyrographite (LDP) (1.35–1.6 g cm^{-3}) in the HIP, their weight in the hydrogen flow at 1,570 K decreases for 6,000 s by approximately 3–8 %, while the strength decreases by 30 %.

The strength of casing from ZrC composition with pyrographite after annealing essentially decreased. Some casings (No 2, 5, 6, Table 6.2) were destroyed at a light touch.

The loss in the mass of cases made of PGV pyrographite with distinctly anisotropic properties (density from 1.2 to 2.25 g cm^{-3}) in hydrogen at 2,300 K for 4,000 s reaches 45 %. Due to a low porosity, the interaction of uncoated pyrolytic graphite with hydrogen, for example, at 2,100 K is two times weaker than that for usual pressed graphite types.

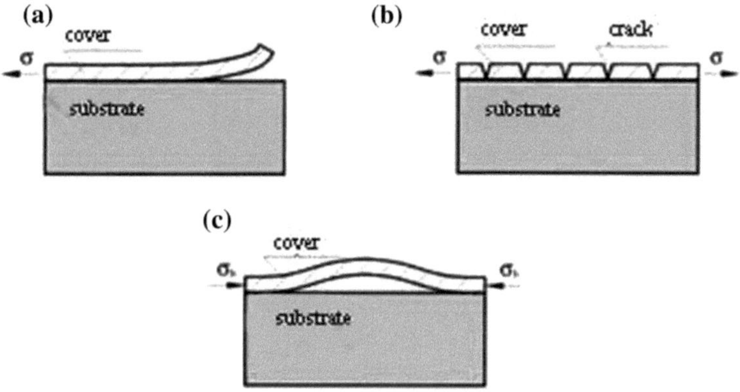

Fig. 6.8 Fracture view of thin coatings under influence of residual stresses: **a** Flaking from tensile stresses at low coating adhesion; **b** cracking fissuring from tensile stresses at good adhesion of a coating with a substrate; **c** buckling coating at formation of compression stresses and its subsequent fracture [6]

Table 6.2 Density and loss of free carbon from the ZrC composition with pyrographite after exposure

		Test condition		C_{free} (% weight)		After test	
No	γ (g/cm^2)	T (K)	τ (s)	Initial	After test	$\Delta C/C_{init}$	$\Delta\gamma/\gamma$
1	3.5	2,270	10,000	25.5	18.3	0.28	0.57
2	3.1	2,270	10,000	19.7	8.9	0.54	0.9
3	3.0	2,470	6,000	34.5	20.9	0.39	1.3
4	3.3	2,470	4,000	19.7	4.2	0.78	0.9
5	3.5	2,470	6,000	27.0	8.9	0.67	0.57
6	2.8	2,470	4,000	22.0	2.7	0.87	0.1

References

1. Daragan, I. D., D'jakov, E. K., Fedik, I. I. et al. (2003). Fuel element assemblages of the space nuclear power propulsion systems. Moscow: Nuclear Technology Engineering Industry, vol. IV-25, under Adamov's edition. Engineering industry, book 2.
2. Nuclear Technology Engineering Industry (2003). Moscow, vol. IV-25, under Adamovs's edition. Engineering industry., book 1. 953 p. book 2. 943 p.
3. Jackson, H. F., & Le, W. E. (2012). *Properties and characteristics of ZrC* (pp. 339–372). Amsterdam: Elsiver.
4. Lutikov, R. A., Savin, V. I., Fedorov, E. M. et al.(1993). *Interacting of some structural and fuel materials with hydrogen* (8 p.). *The Report at Conference: Nuclear Power in Space.*
5. Lanin, A. G., Leonov, S. V., Turchin, V. N. et al. (1987). *Influence of nitrogen and hydrogen gas medium on mechanical characteristics of zirconium and niobium carbides over temperatures range 300–2700 K* (77 p.). Scientific Research Institute "Luch".
6. Lanin, A. G., & Fedik, I. I.(2008). *Thermal stress resistance of materials* (p. 239). Heidelberg: Springer.

Chapter 7
Bearing Capacity of Elements' HGA

7.1 Fracture Criteria of Thermally Loaded Bodies

Temperature stresses inevitably appearing during the NRE operation are one of the main factors that can cause the destruction of ceramic units of HRAs [1]. In some cases, temperature stresses determine the construction features and output parameters of a device being developed and the possibility of using one material or another. Therefore, to estimate the strength, it is necessary to determine the level of these stresses and the degree of their danger for individual elements and the construction as a whole.

As a rule, the strength is calculated by using admissible normal or tangential stresses that are safe for the product strength or the limit number of loading cycles. For a newly created class of machines or apparatuses, it is necessary to determine the properties of the collapse of materials from which their bearing elements are made and to determine certain criteria for the strength and its limiting values. The strength of a metal HRA housing is estimated by special methods for calculating the strength and the radial and longitudinal stability of cylindrical shells known in rocket building. The bearing capacity of ceramic fuel elements with a cross-section of a complicated shape was estimated by using the strength criteria for bodies operating in the inhomogeneous field of thermal stresses developed at the institute Luch.

The results of studies initiated at the institute as early as 1973 showed that the use of methods of thermal action in different combinations and varying the shape and size of a body changed the thermal strength and the type of body collapse [3]. Based on the concepts of force fracture mechanics, a new criterion factor N was introduced [2], which took the stress distribution into account and determined conditions of the total or partial fracture of bodies upon changing this stressed state. The values of the parameter N for different types of thermal loading were calculated numerically.

Upon heating the side surface of a body made of an elastic brittle material, a crack appearing in the central tensile region causes the complete fragmentation of the body when the critical stress intensity K_{1c} and values $N \geq N_{cr}$ are reached (Fig. 7.1a). When the side surface of a heated body is cooled (under conditions similar to the

A. Lanin, *Nuclear Rocket Engine Reactor*, Springer Series in Materials Science 170,
DOI: 10.1007/978-3-642-32430-7_7, © Springer-Verlag Berlin Heidelberg 2013

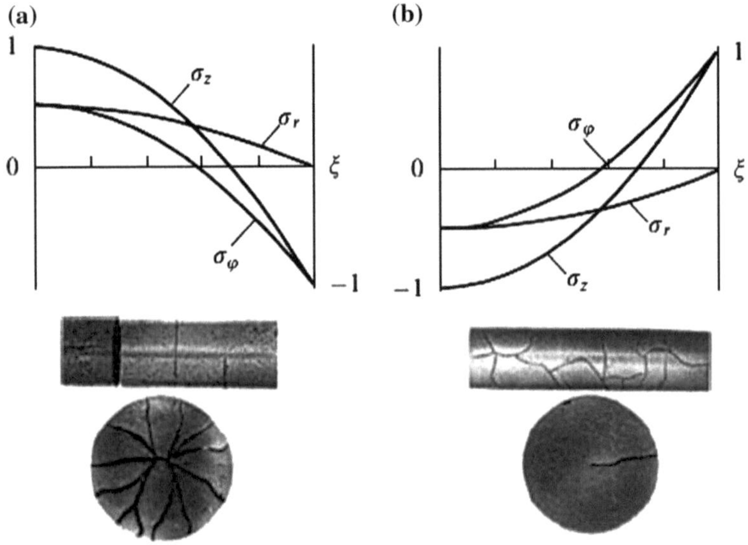

Fig. 7.1 Change in the destruction of ZrC samples from complete fragmentation upon heating the body surface (**a**) to partial destruction caused by surface cracks appearing upon cooling (**b**) due to a change in the stressed state of the body [2]. (Three stress components for a cylindrical sample and the absence of the axial component σ_Z in a thin disc)

operation conditions of HREs according to the scheme presented in Fig. 7.1b) and the inhomogeneity parameter is $N < N_{cr}$, the nonequilibrium propagation of a crack started in the tensile region at the critical value K_{1c} changes to its equilibrium increase. The total fragmentation of the body becomes possible when stresses (after their substantial redistribution) are 8–10 times higher than the start stress of the crack [2]. In this case, the penetration of the crack into the body is 0.55 R on average, which agrees with calculations.

The tests of heated HREs of different compositions by the method of nonstationary cooling in water showed that they were partially damaged due to the appearance of surface cracks at stresses exceeding the tensile stress σ_t corresponding to the heat flow q_s и $2.5\,MW\,m^{-2}$ by only 15–20 % [3]. The bearing capacity of the HRE estimated from decay in the strength decreases almost three times after the appearance of surface cracks and remains virtually unchanged after repeated cyclic loadings. We note that the number of cracks per unit surface increases when q_s increases, while their penetration depth in the body and hence the bending strength do not change (Fig. 7.2). Tests of HREs by passing an electric current through them and blowing off their surface by a gas flow at surface temperatures (1,500–1,900 K) higher than upon cooling in water confirm that ZrC+UC HREs were damaged for $q_s = 2.5{-}3.0\,MW\,m^{-2}$ (Fig. 7.3). Doping a carbide matrix with carbon inclusions almost doubles the damage threshold (up to q_s $5\,MWm^{-2}$). The complete fragmentation

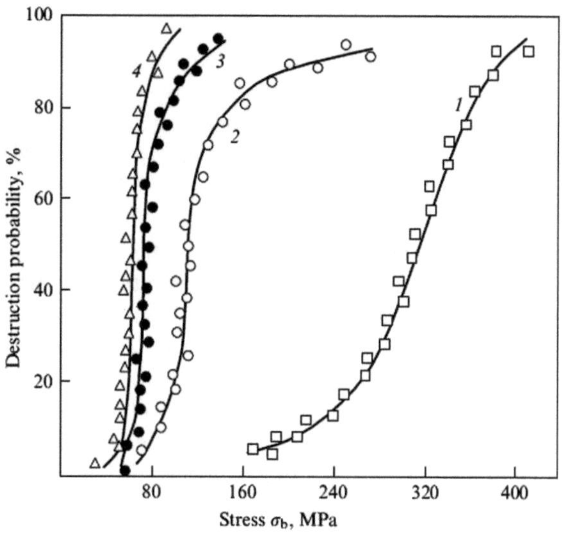

Fig. 7.2 Destruction of the heated $(Zr_{0.97}U_{0.03})C_{0.96}$ fuel elements (up to 570 K), after their nonstationary cooling in cold water, estimated from the decrease in the bending strength from the initial state (*1*) after the first loading (*2*), and after 10 (*3*) and 50 (*4*) cycles

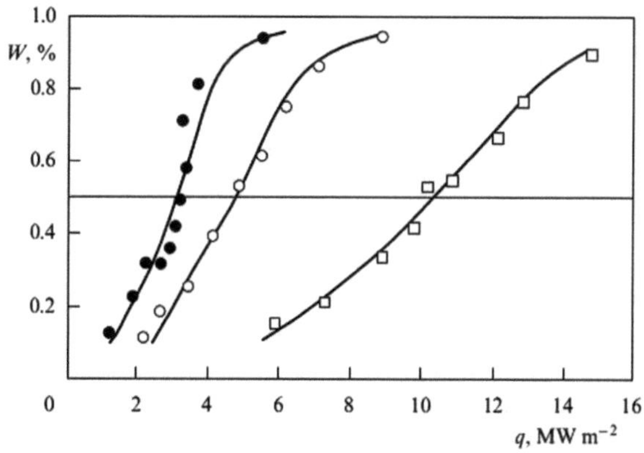

Fig. 7.3 Probabilities W (%) of the destruction of ZrC + UC (•) and ZrC + UC + C (°) fuel elements by surface cracks and of their full fragmentation (□) as functions of the specific thermal flux q at temperatures И 1,500 K [5]

of HREs made of ZrC + UC and ZrC + NbC + UC occurs at mean values $q_s = 10 - 12\,\mathrm{MW\,m^{-2}}$.

The estimation of vibration strength of HRA is made at room temperature without radiation. More than 15th overloading during 50 h in a frequency kilohertz range, and singular blows at the case in three directions did not cause any change of HRA state [4].

The working capacity of fuel elements (FE) and HGA in reactor conditions, modeling modes of reactors' operation of the nuclear propulsion reactor (NPR), and nuclear propulsion energy reactor (NPER) on a propulsion mode (PM) with two various power levels implemented in reactors IVG.1, IR-100 and on an energy mode (EM) in rector RA [6, 7] (Fig. 7.3).

NP tests were carried out in two HGA modifications cooled by hydrogen technological canals 300 (TC-300) and experimental technological canals (ETC) of reactor IVG-1, FE tests implemented in suppressed and filled with helium canals TC-100 of the reactor IR-100 and in filled helium ampoules of reactor RA. In total, 152 technological canals of various type have been tested on PM. Reactor tests of the irradiated FE of various heating sections (HS) of HGA and FE of the resource ampoules were examined. In total, 110 various types of canals after reactor tests were investigated.

Fuel elements of nuclear rocket engine (NRE) were tested in IVG-1 research reactor in the modes that simulate operating conditions of NRE reactor. These fuel elements were used in fuel assemblies (FA), containing several heating sections (HS) 100 mm long inserted into the flowing gas-cooled technological channels of NRE reactor core.

During the working out of NRE fuel elements in IVG-1 reactor, 30 technological channels with 8- or 6-cell fuel assemblies were installed. Each HS in the 8-cell assemblies contains 379 fuel elements, and each HS in the 6-cell assemblies—151 fuel elements (beam diameters of fuel elements in the 8-cell and 6-cell assemblies were respectively 47.0 and 29.7 mm). In the 8-cell assemblies the first five HS contain fuel elements of (Zr, U)C and the last three are fuel elements of (Zr, Nb, U)C; in 6-cell assemblies, the first four HS contain fuel elements of (Zr, U)C+C, and the last two are fuel element of (Zr, Nb, U)C.

During the ground test of NRE fuel elements each IVG-1 reactor start-up (of about 5–6 min duration) corresponded to one of the reactor activation of space NRE. In the trial tests of the IVG-1 reactor (first core) only three start-ups were performed—power start-up (PS) and two working start-ups (WS). And in the life cycle tests of the second core the regulated number of the start-ups was made (that is, 1 PS and 11 WS). At that the modes of WS in the life cycle tests corresponded well to operating conditions of the reactor activations of space NRE. In particular, the hydrogen temperature at FA outlet reached 3,100 K. Hydrogen pressure at the inlet and outlet of FA was $\sim$10 and $\sim$5 MPa, respectively, the maximum temperature drop along the radial cross-section of fuel element reached 250 K.

7.2 Operating Conditions of NRE Fuel Elements

Operating conditions of fuel elements in NRE reactor (and test conditions of NRE fuel elements in IVG-1 reactor) are associated with essential presence of thermal stresses in them, because all the fuel elements heat from the internal energy is selected by the hydrogen flow pumped through the fuel assembly. The most intense heat removal from the surface of the fuel element (creating a radial temperature drop ΔT over its cross-section) takes place in the inlet HS, operating in the temperature range of brittle state that means the first and second HS in the 6-cell FA or the first, second, and third HS in the 8-cell FA.

In the course of the reactor steady-state operation, the temperature drop in each fuel element increases from zero to maximum and remains at that value during the

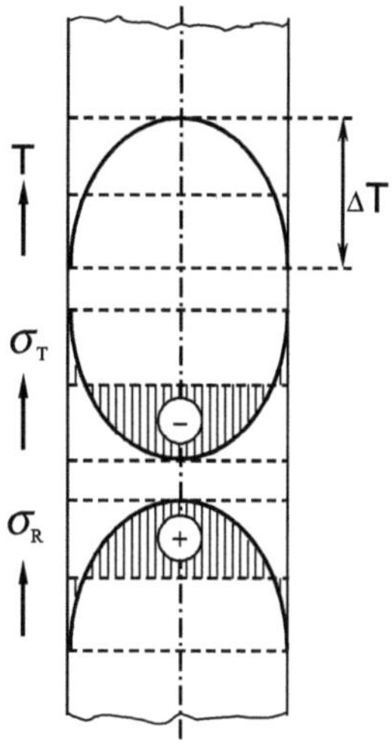

Fig. 7.4 The radial distribution of temperature T, the axial thermal stresses σ_T and axial residual radiation stress σ_R resulting in NRE fuel elements at various stages of its operation in flowing technological channels

start-up (the parabolic form of the temperature distribution in cross-section of fuel element is shown in Fig. 7.4). Temperature drop value is determined by the heat flux q_s from the surface of a fuel element, thermal conductivity λ of the fuel element material and the radius r. This temperature drop ΔT creates dangerous thermal macro stresses σ_T with tensile component at the fuel element surface (see the diagram of axial σ_T in Fig. 7.4). The values of the axial and tangential tensile stresses σ_T can be found [8] from the formula:

$$\sigma_T = \alpha E \Delta T/[2(1-\nu)] = \alpha\ Eq_s r/[4\lambda(1-\nu)], \quad (7.1)$$

where α—coefficient of thermal expansion; E—Young modulus; ν—Poisson's constant.

Thermal stresses σ_T pose a real danger for fuel elements operating only at temperatures up to 1,600 K. Because in these conditions, the brittle refractory carbides will not have relaxation capability of these time elastic macro stresses. If the stress σ_T exceeds the tensile strength of fuel element material there will be cracks on its surface. Cracks in themselves do not affect the operation of the fuel elements, but their appearance leads to very negative consequences, as in the future failure of fuel rods weakened by cracks by bending, bandage and vibration loads (bending loads occur in the fuel elements in the event of the longitudinal beams twisting in HS, and bandage—under the influence of radiation thickening of fuel elements).

Along with the negative aspect of hazardous thermal stresses presented in the fuel elements there is a positive sign consisting of fuel elements hardening due to the appearance of compressive residual radiation macro stresses on their surfaces σ_R (see diagram of axial σ_R in Fig. 7.4). Residual stresses σ_R arise again because of the presence of radial temperature drop ΔT in the irradiated fuel element, which provides radiation swelling heterogeneity of the fuel element material. Due to different rates

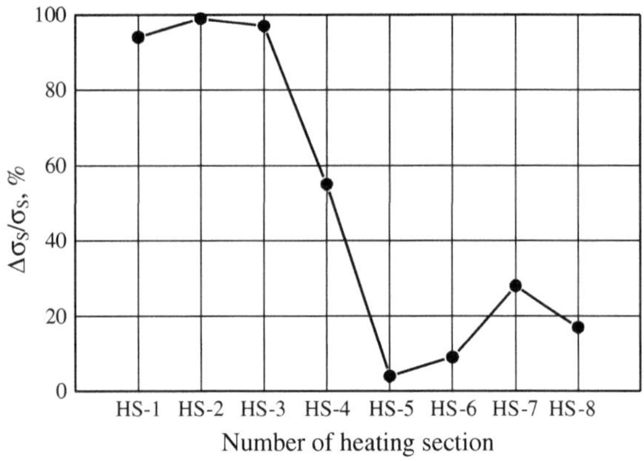

Fig. 7.5 The nature of fuel elements strength change $\Delta\sigma_S/\sigma_S$ on FA length after trial tests in three start-ups PS+2WS of IVG-1 reactor with the first core

of annealing of resulting radiation defects volume of peripheral (cooler) regions of the fuel element will increase stronger than the internal volume of (hotter) regions. This will ensure the appearance of compressive residual stresses on the fuel element surface and tensile stresses inside, as the internal regions preventing excessive expansion of the surface regions, will create an elastic compression in them.

7.3 Trial Tests of NRE Fuel Elements

Trial and life cycle tests of NRE fuel elements in flowing gas-cooled channels of IVG-1 research reactor showed that the state of the irradiated fuel elements to a great extent depends on the specific modes of reactor tests [9–11]. Thus, the fuel elements of (Zr, U)C and (Zr, Nb, U)C after the trial tests in the channels with 8-cell fuel assemblies in the modes, significantly reduced compared to the nominal mode for NRE reactor, were discovered in a quite satisfactory condition: surface cracks and fracture of the fuel rods were few events, the longitudinal beams twisting of fuel elements in HS were small, the increase in the coefficient of gas-dynamic resistance of the cooling duct of fuel assemblies was also small. At the same time a significant post-irradiation increase of strength σ_S and thermal strength of fuel elements, as well as the presence of compressive residual stresses σ_R on their surfaces was recorded.

Hardening of fuel elements $\Delta\sigma_S/\sigma_S$ in three inlet (low-temperature) sections (HS-1, HS-2, and HS-3) reached almost 100 %, in four outlet (high-temperature) sections it was small (Fig. 7.5).

The nature of fuel elements thermal strength increase was the same as the character of their hardening. Measurements of fuel elements thermal strength by time-dependent thermal loading (destructive temperature drop ΔT_F or the first criterion of the thermal strength was determined by rapid immersion of the pre-heated fuel rods in water) showed the more than twofold increase in thermal strength of fuel elements of (Zr, U)C in the first three inlet HS and the almost complete no change of

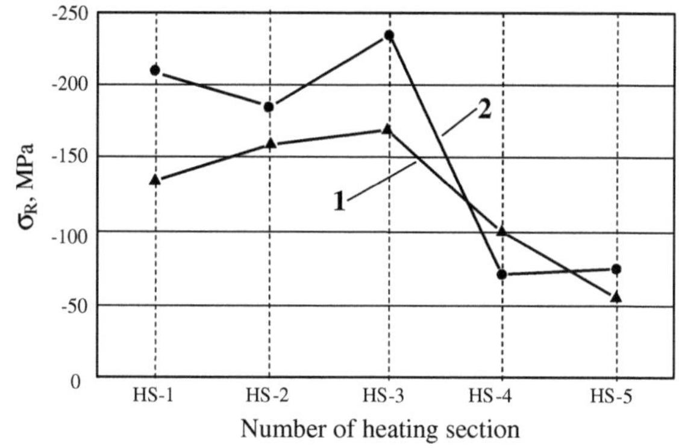

Fig. 7.6 Residual stresses σ_R on the surfaces of fuel elements of (Zr, U)C in the first five HS of 8-cell FA after tests in one start-up WS (*1*) to ND burn-up $0.9 \cdot 10^{17}$ div/cm^3 and three start-ups PS+2WS (*2*) up to ND burn-up $2.3 \cdot 10^{17}$ div/cm^3

the characteristics of fuel elements of (Zr, U)C and (Zr, Nb, U)C in high-temperature HS. The coincidence of the change of the strength and thermal strength of fuel elements on FA length is due to the fact that the thermal strength of brittle materials products is linearly related to their strength.

The results of X-ray measurements of the magnitude and sign of axial residual stresses σ_R in fuel elements of two channels exposed to different uranium burn-up ND (ND burn-up—the number of uranium fission per a volume unit) are shown in Fig. 7.6. In the measurements, we used the developed nondestructive method [12] with photo recording of the stresses in a special X-ray camera. The measurements were made on the fuel elements which had no surface cracks. When comparing the data in Figs. 7.6 and 7.5, we can see the satisfactory correlation of the change in hardening levels of fuel elements along the FA length and the levels of compressive residual stresses in them: the maximal hardening $\Delta\sigma_S/\sigma_S$ and maximum stresses σ_R occur in the fuel elements of the first three HS of both channels.

Furthermore, using only the fuel elements from HS-1 sections of six channels. X-ray study of the uranium burn-up effect on the level of occurred stresses σ_R was conducted out. It turned out that ND increase on an order (from $\sim 2.1 \cdot 10^{16}$ to $\sim 2.1 \cdot 10^{17}$ div/cm^3) leads to the more than twofold increase of stresses level σ_R. This has led to express concern for the integrity of the fuel rods involved in a longer state life tests at IVG-1 reactor with the second core. Indeed, a further significant increase of ND burn-up the positive effect of increasing compressive stresses at the surfaces of fuel elements may be replaced by the negative: a significant increase of the tensile axial stresses σ_R inside of fuel elements could lead to their damage into fragments at the time of cooling after start-up (at the time of temporary stresses disappearance σ_T).

The question on the danger of expected excessive growth of stresses σ_R in the fuel elements was resolved only after the completion of their life cycle tests in the second IVG-1 reactor core, when the opportunity to X-ray measurements of stresses σ_R in fuel elements with ND burn-up in a wide range (from $\sim 2 \cdot 10^{16}$ to $\sim 2 \cdot 10^{18}$ div/cm^3). To do this, in addition to data on the fuel elements of (Zr, U)C tested in the six channels of IVG-1 reactor first core, X-ray diffractometry parameter measurements

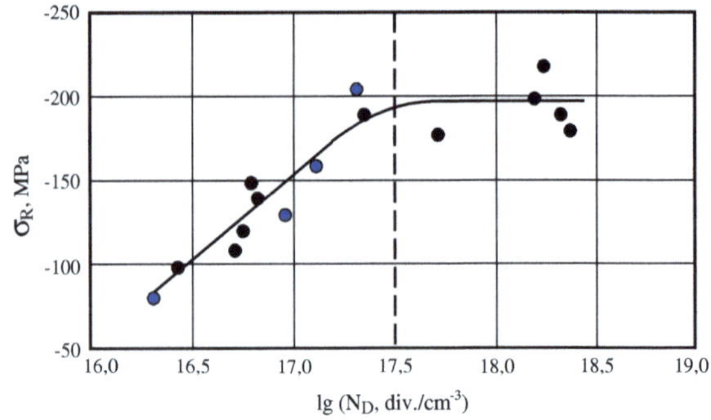

Fig. 7.7 The levels of residual stresses σ_R on the surfaces of fuel elements of (Zr, U)C + C and (Zr, U)C depending on ND burn-up in the range of $\sim 2 \cdot 10^{16} - \sim 2 \cdot 10^{18}$ div/sm^3

of the magnitudes and signs of axial residual stresses σ_R on the surfaces of fuel elements of (Zr, U)C+C tested in low-temperature HS of 11 channels of IVG-1 reactor second core were performed. The measurements were carried out on the selected integer fuel elements containing no surface cracks.

The results of measurements σ_R (see Fig. 7.7 in which data for the first core fuel elements are marked with square dots) indicated that the continuous growth of the radiation dose up to uranium burn-up about $2{\cdot}10^{18}$ div/cm^3 was not accompanied by continuous growth of compressive stresses σ_R on the surface of a fuel element (and tensile σ_R inside) to excessively high values. The increase of stresses σ_R (fuel element material swelling heterogeneity increase), as seen from the figure was stopped after burning of about $3{\cdot}10^{17}$ div/cm^3 (indicated by dotted line) before reaching 200 MPa, that eliminates the risk of failure of fuel elements from the tensile residual stresses operating in their internal regions.

In contrast to the first core of IVG-1 reactor, the second core fuel elements tested in modes close to the nominal operation of NRE reactor, there was a significant degradation of the initial state, especially noticeable in low-temperature HS-1 and HS-2. In particular, the fuel elements in these sections had numerous cracks and damage, and the longitudinal beams twisting in HS and the numbers of failures of fuel elements were increased with increasing the number of reactor start-ups. For example, after six start-ups the angles of beams twisting of fuel elements reached about ~20°, and the number of broken fuel elements (Fig. 7.8) was approximately 80 %.

Destroyed fuel elements in HS-1 and HS-2 were small (3–15 mm in length) and slightly mixed fragments of rods that significantly—hundreds of percent—increased the coefficient of gas-dynamic resistance at the entrance of FA cooling duct. The fuel elements in the remaining four high-temperature HS remained intact, or were damaged only to large (30 mm length) fragments without mutual displacement, their strength decreased with increasing temperature and duration of exposure due to the degradation of the surface material by erosion and hydration.

It is obvious that the original cause of the negative situation in the inlet HS is the lack of thermal strength of carbide-graphite fuel elements at the temperature of the brittle state. Indeed, the level of thermal strength of these fuel elements is such

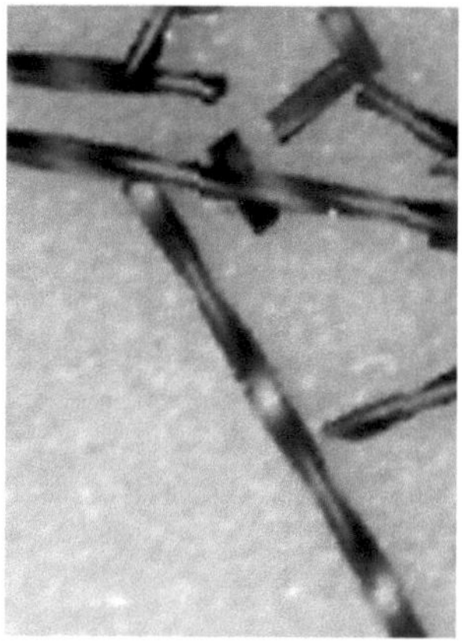

Fig. 7.8 View of fuel elements fracture from the second heating section (HS) (integrated flux of fast neutrons of order $10^{19} sm^{-2}$)

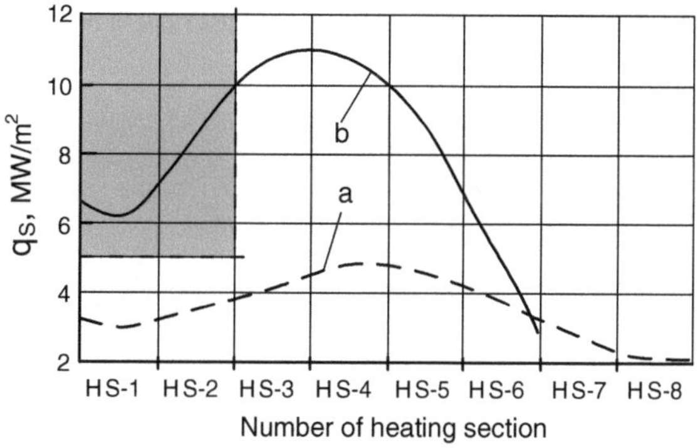

Fig. 7.9 Distribution of heat fluxes from the surface of fuel elements q_S along the length of 8-cell (**a**) and 6-cell (**b**) FA, respectively, in trial and life cycle tests of fuel elements in IVG-1 reactor

that the cracks appear in them even at heat loads q_S higher ~5 MW/m^2 (see upper dangerous region in Fig. 7.9). That is why the fuel elements of the first two sections of 6-cell FA were damaged by cracks and destroyed during tests in IVG-1 reactor second core, and the fuel elements in the first three sections of 8-cell assemblies remained intact during the tests in the first core.

The cracks arising in FE under the influence of thermoelastic stress lead to a final fracture in the first and second HS by a tightness of ten-order microns on a series of the peripheral mainly damaged FE. Generated FE fragments can drop out and increase hydraulic resistance of HS, in a case if they have moving freedom in any direction. It depends on a concrete locating of fragments and also on angular orientation surrounding the yielded FE fragments and forces, capable to move them (Fig. 7.10).

Fuel elements in the last HS undergo plasticity at elevated temperatures, forming sometimes through axial channels (Fig. 7.11). The decline of the heat removal, due to change of gas flow in the channel, leads to temperature increase of fuel elements that in the presence of compression loading promotes emersion of plastic deformation.

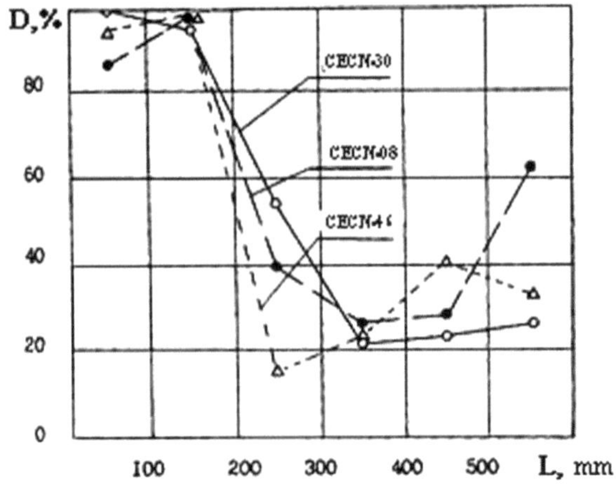

Fig. 7.10 Quantity of fuel element damage D (%) in heating sections of the central experimental channel (CEC) of HGA

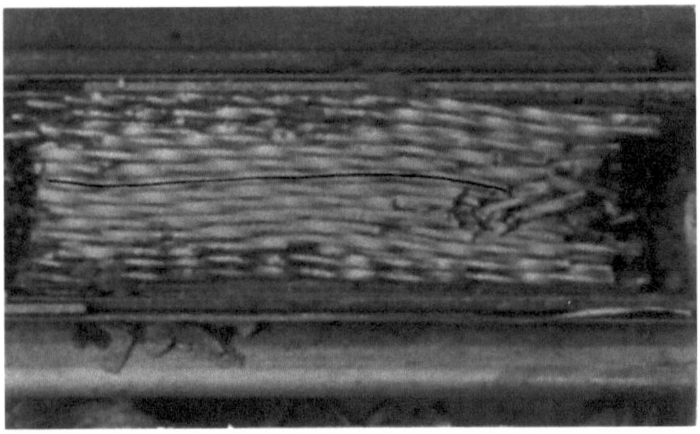

Fig. 7.11 Plastic strain of the fuel elements of the sixth HS

Fig. 7.12 Twist of the fourth and fifth heating sections of the central HGA

FE of the last high-temperature HS can also be twisted around axis and lose the stability in the presence of axial compression load under the influence of axial forces of the downthrusted spring and power loads, arising at gas passing in HS (Fig. 7.12).

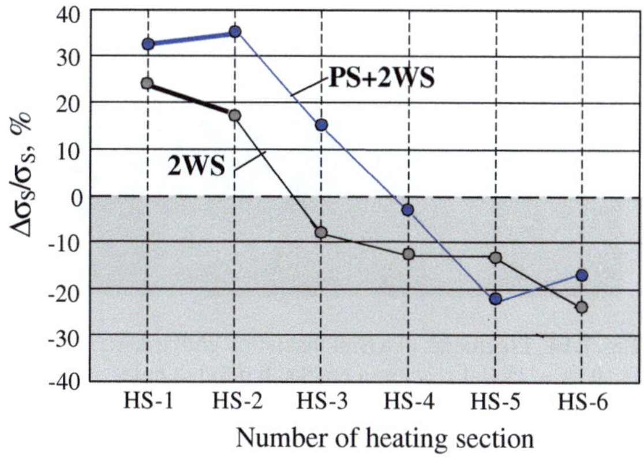

Fig. 7.13 Change of fuel elements strength $\Delta\sigma_S/\sigma_S$ on FA length of two channels tested in the fourth core of IVG-1 reactor

Consequently, for the elimination of a failure of fuel elements in the first two sections of 6-cell assemblies and prevent processes of grinding and mixing of the fragments it is necessary, as seen in Fig. 7.9, to increase the initial thermal strength of the fuel elements of (U, Zr)C+C approximately double. As the level of thermal strength of brittle carbide material can be elevated by increasing its strength, in [10] the possible ways of pre-hardening of fuel elements by thermal, radiation, and combination methods were analyzed. Among the possible radiation methods, the way of fuel elements strengthening by residual radiation stresses σ_R, in which the fuel elements of 6-cell FA were previously irradiated to ND burn-up about $3{\cdot}10^{17}$ div/sm^3 in a mode of decreased power (at average heat load on FA length q_S about 4 MW/m^2), was pointed to. Pre-irradiation is carried out in the shape of the first start-up (FS) of NRE space facility located in orbit. Double hardening of carbide-graphite fuel elements of low-temperature HS achieved during FS must ensure the absence of cracks and breakage in the subsequent reactor activations.

A comparison of the change of fuel elements strength after reactor tests in only a few (n) working start-ups (nWS) and after the tests in the n working start-ups with prior power start-up (PS+nWS) [10, 11] can serve as an experimental substantiation of the proposed method.

Such positive effect of PS start-up was recorded among the fuel elements tested in the IVG-1 reactor fourth core (see Fig. 7.13): hardening of fuel elements in HS-1 and HS-2 after the tests in $P_S + 2W_S$ and $2W_S$ amounted to ~35 and ~20 %, correspondingly.

7.4 Bearing Capacity of the HIP

Reactor tests of the HIP, in general, confirmed the results of laboratory tests on the electron arc installation (EAI). The traces of interacting working medium with the surfaces of the TIP from low density graphite (LDG) were detected. Intensity of

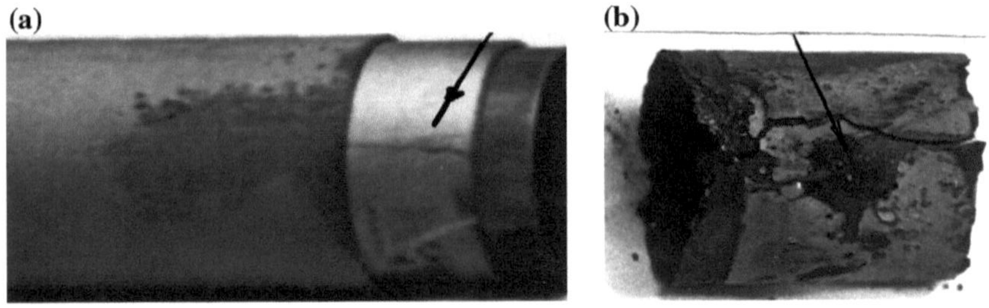

Fig. 7.14 Fragments of a heat insulation with traces of material ablation after nuclear reactor tests in HGA of IVG-1 reactor; **a** cracks, **b** through holes and layers separation

hydrogen interaction increased with growth of temperature along gas traffic route. The fourth heating section (HS) was observed sometimes through holes in HIP. Material ablation in these HIPs made up an order of 25–50 weight%. The most defective was outer range of TIP from high density graphite (HDG).

Casings from ZrC with P_q flaking forming thermal insulation in high temperature part of HRA, in general, kept the geometry though the casings have some cracks, and at dismantling they scattered on separate fragments. The carbide casings also undergo some changes. There were separate cracks and grids of cracks. The overhead part of the HIP near to the supporting exhaust knot is underwent changes also (Fig. 7.14).

Thus, the basic defects of HIP were the cracks and an area of material flushing. The last is connected, both with chemical interacting and with an erosive material removal. The most dangerous from the point of working capacity is the ablation of HIP under contact of hot working medium with bared metal casing. The structure imperfection of HIP is increased with operating time. There is great scattering between alternative technological versions of HIP preparation. More optimum materials' composition and reliable design of HIP will be chosen in the subsequent time of NRE development.

7.5 Bearing Capacity of a Bearing Grid of the HRE

Damage of the bearing grid in the course of reactor operation and loss of bearing capacity may lead to an emergency of the reactor. Damage of bearing grids (BG) was revealed at dismantling of channels of reactor IVG-1 after various start-ups in the aspect of linings' separation from adjoining rows of rods. We remind that the BG are made in the form of sintered block of twisted—four blade rods from ZrC, NbC carbides equimolar composition soldered to lateral surfaces of the block segment facings from pure carbide ZrC.

Though the load-bearing capacity of support grids (SG) was retained, the fact of its fracture confirmed an uncontrollable reliability loss of reactor operation. This necessitated the analysis of the stress fields of SG during operation and the need for deeper understanding of possible fracture mechanisms in order to develop recommendations

Table 7.1 Possible strengthening methods for rod-shaped fuel elements

	Designated method	$\Delta\,\sigma/\sigma$ (%)
1	Thermal cycling in temperature range from TBP to 2,700 K	20
2	Quench hardening by thermal radiation cooling from 2,700 K	~30
3	Quench hardening by forced cooling from 2,700 K	~60
4	Thermal healing of initial cracks at 2,500 K	~40
5	Reactor radiation creation of surface compressive stresses	~80
	Combined method 2.4.5	100

for improving SG performance [13]. The results of stress analysis show that SG may be fractured during unstable start stage (before reaching steady-state operation mode) or, under certain conditions, also during cooling stage. When approaching the steady-state regime highest thermal stresses may be reached: up to 400+MPa for tangential tensile stresses on the SG periphery, and up to −70 MPa for compressive stresses near the SG center. Under such stresses, segment linings may crack and delaminate from soldered rod block.

Massive segment linings are proposed to be replaced with thin-walled split-type casing with semi-cylindrical fillers to avoid generation of dangerously high residual stresses. The splitting angle depends primarily on the prerequisite of retaining the rods in a package during assembly and subsequent processing. Angle value may fluctuate over a wide range: from 5 to ~90° relative to generating line. Implementation of such casing for grid assembly presents no problems, while being also less labor-intensive as compared to lining design. Testing of an experimental batch of proposed SG design has confirmed triple increase of thermal stability of split-type casing SG, as compared to lining design.

7.6 Possible Methods for Increasing Bearing Capacity of Ceramics

Improving of bearing capacity of ceramics with a limited plasticity in a broad temperature range should be done taking into account operation parameters. For fuel elements of the first sections of the HRA operating in the brittle-damage temperature range, it is necessary first of all to decrease the defectiveness of ceramic using various hardening methods capable to elevate the strength. Optimizing hardening methods by modification of stressed surface state the strength of elements can be enhanced up to 100 % (Table 7.1).

The optimal choice of the parameters of traditional isothermal sintering of ZrC samples in different media (hydrogen, argon, and vacuum) allows achieving a high density (not less than 95 %) and the bending strength about 550 MPa at sintering temperatures 2,500–2,700 K [2, 14].

References

1. Fedik, I. I., Kolesov, V. S., & Mihajlov, V. N. (1985). Temperature fields and thermal stresses in nuclear-reactors (280 p.). M. Energoatomizdat.
2. Lanin. A. G., & Fedik, I. I. (2008). Thermal stress resistance of materials (p. 239). Heidelberg: Springer.
3. Lanin, A. G. (1998). Strength and thermal stress resistance of structural ceramic (111 p.). -M. Moscow State Engineer Physical institute.
4. Nuclear Technology Engineering Industry, Moscow, vol. IV-25, under Adamovs's edition. Engineering industry., book 1. 953 p. book 2. 943 p.
5. Daragan, I. D., D'jakov, E. K., Fedik, I. I. et al. (2003). Fuel element assemblages of the space nuclear power propulsion systems. Moscow: Nuclear Technology Engineering Industry, vol. IV-25, under Adamov's edition. Engineering industry, book 2.
6. Zaharkin, V. I., Ionkin, V. A., Konovalov, et al. Working out of NRER on the basis of the reactor of minimum sizes IRGIT. Power tests of a developmental type of a nuclear reactor (pp. 271–278). Third Branch Conference "Nuclear Power in Space".
7. Degtyaryova, L. S., Parshin, N. J., & Popov, E. B. (2003). Research of geometrical and hydraulic characteristics in fuel elements sections of NRER (pp. 17–21). Collection of works of Tsniiatominform.
8. Vlasov, N. M., & Fedik, I. I. (2001). Fuel elements of nuclear rocket engines (p. 207). Moscow: Tsniatominform.
9. Derjavko, I. I., Zelenskij, D. I., Lanin, A. G. et al. (1999). Reactor working off of the rod carbide fuels of NRER (pp. 539–543). 5th International Conference "Nuclear Power in Space", Podolsk.
10. Deryavko, I. I., Chernyadev, V. V., Gorin, N. V., et al. (2010). Characteristics of identifying the fuel rods of NRE. Vestnik NNC RK. edn.,4(44), 95–105.
11. Deryavko, I. I. (2004). Serviceability of rod ceramic fuel elements on engine modes of NRE reactor. Vestnik NNC RK. edn.,4(20), 90–98.
12. Derjavko, I. I., Egorov, V. S., Lanin, A. G. et al. (2001). Radiographic research of the residual Stresses in the rod carbide fuel elements. The Bulletin of the National Nuclear Centre of Republic Kazakhstan, vol. 14. p. 95.
13. Ponomarev-Stepnoy, N. N., Rachuk, V. S., Smetannikov. V. P., & Fedik, I. I. (2005). Space nuclear power and power propulsion systems on the basis of the reactor with external transformation of heat in a hard atomic zone (pp. 45–51). The International Conference "Nuclear Power in Space-2005".
14. Bulychyov, V. P., Andrievsky, R. A., & Nezhevenko, L. B. (1977). Sintering of zirconium carbide. *Powder Metallurgy, 1*(4), 38–42.

Chapter 8
Outlook for Nuclear Rocket Engine Reactors

NRE reactors of different Nuclear Engine Power Installations (NEPI) based on tested technologies have actually no alternatives in deep space investigations with the help of unmanned space probes and piloted interplanetary devices [1]. The developments of NREs were stopped only temporarily because humankind could not afford to spend a huge amount of money for large-scale cosmic studies. It is most likely that the NRE program should become an international collaboration program in the future, like programs for the development of the international thermonuclear experimental reactor (ITER), high-power accelerators, etc. The program would have a global character, in view of its legal. Ecological and other aspects, because the program involves the use of fission materials of the highest armament quality, it should be under international control. NRE reactors for different purposes would be based on different HRE geometries with fuel compositions providing the required efficiency. To fulfill the tasks in cosmic flights, the NRE reactor should have high reliability, with the no-failure probability no less than 0.99.

The different versions of a device operating in many regimes and capable of producing, along with the reactive thrust, the electric energy for ensuring the activity of a spacecraft were extensively developed beginning from the early 1980s [2].

They should work not only in the basic, engine regime but also in two energy regimes at low power for a few years and high power during approximately half of the specified time resource of the engine regime. The high-power regime presents no difficulties in tests because, according to all its parameters, the reactor is loaded much more weakly than in the basic regime of the IVG-1. In the low-power regime, the heat carrier flows around the HRA only outside its housing, while the heat from HREs is transferred to the housing by radiation through heat insulation. This regime is quite different from the basic regime, when the fuel is considerably burned out (down to no less than 3–5 % of the initial amount) and the chemical composition can change due to the incongruent evaporation of materials. Therefore, the efficiency of the HRA and HRE components under these conditions requires additional investigations.

It is clear that the outlook for the development and building of active cores for nuclear space energy devices will be first and foremost related to the improvement

A. Lanin, *Nuclear Rocket Engine Reactor*, Springer Series in Materials Science 170,
DOI: 10.1007/978-3-642-32430-7_8, © Springer-Verlag Berlin Heidelberg 2013

Table 8.1 Exploitation parameter of the NEPI

Parameter	NEPI type NEPI-1	NEPI-2	NEPI-3
Thermal power			
On thrust regime (KW)	950	5,100	5,100
On energetic regime (KW)	220	135	50
Temperature of hydrogen (K)	Near 2,100	2,800	2,800
Thrust force (m/s.)	7,550	8,825	8,825
Exploitations duration			
On thrust regime (h)	250	100	100
On energetic regime (year)	7–10	10	10

in the construction and technology of heating sections containing fuel compositions with the heat-releasing density 40 MW l^{-1} at elevated temperatures (above 3,200 K) in the engine regime and to ensuring that fission products are kept in HREs for a few years at the temperature 2,000 K in a deep vacuum or in hydrogen-containing working substances at pressures from 0.1 to 0.2 atm.

The available design and technological groundwork for working of NRE reactor testify to basic possibility of creation HRA, efficient both on regimes of the reactive thrust, and on regimes of long work in Brighton cycle at a power production for orbiting spacecraft equipment [3]. Over the previous years, a row of thermal energy transformation concepts in electric by bimodal [4] by a fast-neutron reactor with a lithium cooling contour or with thermal emission (accordingly NEED-2 and NEED-3) have been developed (Table 8.1).

Quite probably, for that the raise of working capacity of constructive elements of the NEED engineering production of nanosize materials will be developed [5, 6]. New heat insulating materials, resistant in the hydrogen environment on the basis of fibrous, porous, and multilayered carbides and nitrides of refractory metals will be created.

The Commission of the President of Russia on the Modernization and Technological Development of Russian Economics in 2009 recommended reconsidering the question of developing a spacecraft with a nuclear rocket engine. In 2010, the Government of Russia provided the initial financial support for the development of an outline for a project of a megawatt nuclear energy device, with the possible beginning of the module construction in 2018.

References

1. Mihajlov, V. N., Ponomarev-Stepnoj, N. N., Fedik, I. I. et al. (2005). Prospects of nuclear space energy use in the next century (pp. 11–17). *International Conferences "Nuclear Power in Space—2005", Moscow, Podolsk.*

2. Daragan, I. D., D'jakov, E. K., Fedik I. I. et al. (2003). Fuel element assemblages of the space nuclear power propulsion systems. Moscow: Nuclear Technology Engineering Industry. vol. IV-25, under Adamov's edition. Engineering industry, book 2.
3. Barinov, S. V., Daragan, I. D., Kaminskij, A. S. et al. (2005). Concept of the interplanetary atomic power station for development of Mars on the basses of high-temperature reactor with turbo-generator energy transformation on a gas cycle of Brighton (pp. 498–504). Moscow, Podolsk: Nuclear Power in Space.
4. Demjanko, J. G. Konuhov, G. A., Koroteev, A. S., Kuzmin, E. P., & Pavelev, A. A. (2001). *Nuclear rocket engines* (p. 413). Moscow: Open Company "Norm-inform" .
5. Koch, C. C. Ovud'ko, I. A., Seal, S., & Veprek, S. (2007). *Structural nanocrystalline materials: Fundamental and application* (p. 364). Cambridge: Cambridge University Press.
6. Andrievsky R. A. (2012). *The basis of nanostructural material science. Possibilities and problems* (p. 251). Moscow: Publishing House BINOM.

Index

DOI: 10.1007/978-3-642-32430-7,

MIX
Papier aus verantwortungsvollen Quellen
Paper from responsible sources
FSC® C105338

If you have any concerns about our products,
you can contact us on
ProductSafety@springernature.com

In case Publisher is established outside the EU,
the EU authorized representative is:
Springer Nature Customer Service Center GmbH
Europaplatz 3, 69115 Heidelberg, Germany

Printed by Libri Plureos GmbH
in Hamburg, Germany